D1277608

Concurrent Engineering

Other Books in the Product Development and Design Series

Quality Function Deployment: Integrating Customer Requirements into Product Design
Yoji Akao

Function Analysis: Systematic Improvement of Quality and Performance
Kaneo Akiyama

TQM for Technical Groups: Total Quality Principles for Product Development (forthcoming)
Kiyoshi Uchimaru
Susumu Okamoto
Bunteru Kurahara

Variety Reduction Program: A Production Strategy for Product Diversification
Toshio Suzue
Akira Kohdate

And two related books from other series:

Hoshin Kanri: Policy Deployment for Successful TQC
Yoji Akao

Equipment Planning for TPM: Maintenance Prevention Design
Fumio Gotoh

Concurrent Engineering

Shortening Lead Times, Raising Quality, and Lowering Costs

John R. Hartley

Publisher's Message by
Norman Bodek
President, Productivity, Inc.

Productivity Press
PORTLAND, OREGON

Productivity Press
P.O. Box 13390
Portland, OR 97213-0390
Telephone: (503) 235-0600
Telefax: (503) 235-0909

Cover design: Hannus Design Associates
Printed and bound by: Edwards Brothers
Printed in the United States of America
Printed on acid-free paper

Library of Congress Cataloging-in-Publication Data

Hartley, John (John R.)
Concurrent engineering : shortening lead times, raising quality, lowering costs / John R. Hartley.
p. cm.
Includes bibliographical references and index.
ISBN 1-56327-006-4 (alk. paper)
1. Production engineering. 2. Production management. 3. New products--Management. I. Title.
TS176.H37 1992
658.5--dc20 91-29936
CIP

97 96 95 94 10 9 8 7 6 5 4 3 2

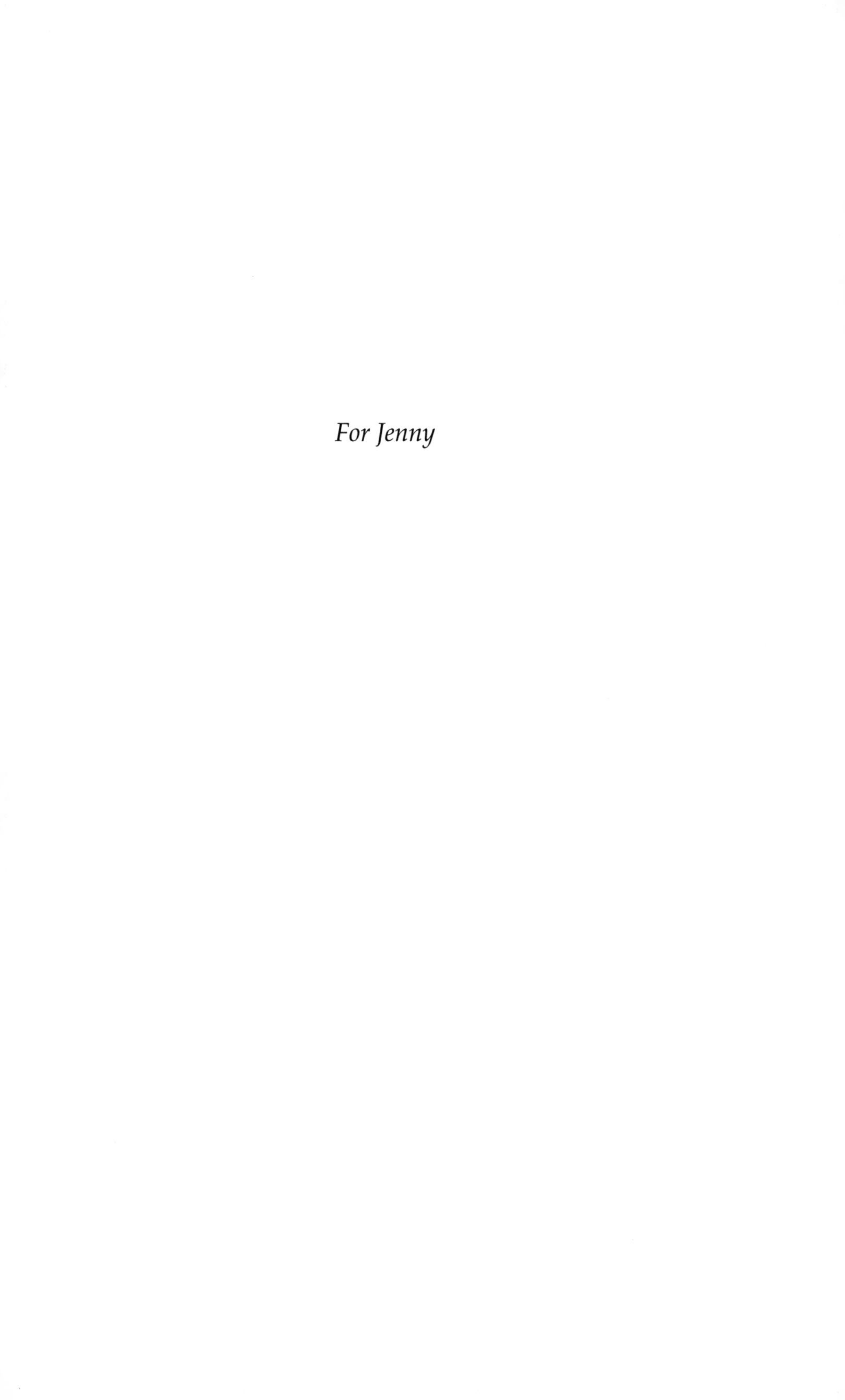

For Jenny

Contents

Illustrations

Publisher's Message

It is now widely believed that U.S. industry's extended period of world dominance in product design, manufacturing innovation, process engineering, productivity, and market share has ended.

The above quote begins a recent study of the National Research Council examining the state of American manufacturing competitiveness. The study goes on to say: "A crucial factor [in this decline] that is not often recognized is the quality of engineering design in U.S. industry. Engineering design is the key technical ingredient in the product realization process. . . . The ability to develop new products of high quality and low cost that meet the customer needs is essential to increasing profitability and national competitiveness."*

Concurrent engineering is far more than the latest fad in manufacturing methodologies. It is the factor, long ignored, where true leverage for competitiveness lies. In discussing concurrent engineering with our Japanese counterparts, Productivity Press was told two significant things: 1) most companies in Japan don't even have a name for this approach, it has been so embedded in the company culture for so long;

* National Research Council. *Improving Engineering Design: Designing for Competitive Advantage*. Washington, D.C.: National Academy Press, 1991, p. 5.

and 2) there is a reluctance to share details about this approach because as the key to shortening time-to-market, concurrent engineering is believed to be the critical factor in competitive leadership for Japan!

To understand this reaction we must consider a few facts that underscore the leverage for quality, cost, and delivery (QCD) that exists in the product design phase. First, the largest percentage of manufacturing costs other than for materials comes from redesign. Thus, it turns out that even if you were to double design costs, for example, from 7 percent to 15 percent of the total product cycle cost, you could reduce the total product cost by 60 percent with concurrent engineering. Second, in a concurrent engineering environment, time-to-market can be reduced by 40 percent or more even if it requires that you double the design time itself. And third, because customer requirements are built in to the design from the start, product quality can now match the needs of the marketplace, at the right time and for the right cost.

Furthermore, without CE, JIT, the QC tools, and TPM remain isolated and limited in their effect to improve a company's profitability. With CE, these approaches are enabled to drive profitability upward and establish a company's competitive strength for the long term. CE creates an environment in which the whole company participates in quality design for the customer.

The concurrent engineering environment results from a total restructuring of the organization. Causing such change may seem like a nightmare for most top managers confronted with this challenge: if you don't do it you don't stand a chance of succeeding through the turn of the century; if you choose to do it, you have to put your organization through a change so major it seems outrageous to consider. And when you are done, your own position of authority over product design may be threatened, or so it seems to many.

These are the issues John Hartley addresses in this insightful book for managers considering implementation of concurrent engineering. The book is full of examples of CE implementation in Japan, Europe, and the United States. Hartley shows how the tools and methodologies — such as QFD, DFM/A, FMEA, 7QC Tools — become integrated and enabled in the CE environment. He also demonstrates the role of CAD and CIM in a concurrent engineering approach, underscoring the necessity to eliminate waste in the process *before* moving to computer integration.

The company examples reveal an extraordinary variety of approaches to implementation of this critical method of product development. Each product type and every company culture may design its concurrent engineering environment differently to suit its strategy and people. Once you understand what it is and what it is meant to accomplish, it is up to you to discover how to create a concurrent engineering environment that works for you. John Hartley's book is an important book for all of our readers. It is especially valuable if you have been looking for some clearly spelled out answers to the gnawing questions that arise when considering implementing this powerful but challenging approach to product development.

We at Productivity Press are proud to offer to our readers this latest volume in our growing Product Development and Design Series. We are grateful to the author, John Hartley, a fine technical writer whose revised version and additional chapters enhance the book's value to American readers; to John Mortimer, editor of the original edition published in 1990; to Industrial Newsletters Ltd., Dunstable, for permission to publish this American edition; and to Malcolm Jones, head of Productivity-Europe, for finding this book and for introducing us to Mr. Hartley and Mr. Mortimer and their publisher.

In addition, we thank all those who worked so hard to produce this high-quality edition: Dorothy Lohmann for managing the manuscript preparation process, Gayle Joyce for managing the production staff; Susan Cobb, Jane Donovan, and Daniel Rabone of the production staff; and Hannus Design Associates for the cover design.

Norman Bodek
CEO
Productivity Press

Diane Asay
Series Editor
Productivity Press

Acknowledgments

The author would like to express his thanks to the following companies for their help in the preparation of this book: Adam Opel, BMW, Comau, Computervision, Digital Equipment, GM Cadillac Division, Hawtal Whiting Group, Honda Motor Co., John Brown Automation, Lamb Technicon, Lucas Automotive, Mazda Motors, Mitsubishi Motor Corp., Nissan Motor, Xerox, Rover Group, Toyota Motor Co., and Volkswagen Audi Group.

Also, he wishes to thank the following companies, individuals, and organizations for permission to reproduce copyrighted illustrations: Computervision (Figures 11-3, 11-6); Digital Equipment (Figures 1-2, 2-3, 4-1, 4-2, 6-2, 17-1); Dowty Group PLC (Figure 2-4); Flex Technologies Inc. (Figures 10-3, 10-4); Harvard Business School (Figure 1-1); Hawtal Whiting Inc. (Figures 2-1, 6-1); Ingersoll Milling Machine (Figures 13-2, 13-3, 13-4); Nissan Motor Co. Ltd. (Figures 3-2, 3-5, 3-6); Rank Xerox (Figures 8-2, 8-3); and IFS Ltd. for permission to publish Figure 8-1, reproduced from *Design for Assembly*, by M. M. Andreasen, S. Kahler, and T. Lund.

He would also like to express appreciation to ISATA for permission to publish a number of figures from papers in the *Proceedings* (Volume 1) *of the 21st International Symposium on Automotive Technology & Automation* (ISATA) given on November 6-10, 1989 in Wiesbaden, Germany. The papers were authored by W. Worreschk of Adam Opel AG, Germany (Figures 5-2, 12-2, and 12-3); R.W. MacDow of Prime Computer Inc., USA (Figure 11-2); and A.W. Aswad and J.W. Knight of the University of Michigan, Dearborn (Figures 3-1 and 3-3). Messrs J.C. Ford, chief engineer, Lucas Car Brake Systems, Pontypool, UK (Figure 14-3) and Dr. I. A. Wulf and R. Sterbl, Control Data GmbH, Frankfurt, Germany (Figures 11-1, 11-4 and 11-5) for figures from the seminars of Autotech '89 sponsored by the Council of the Institution of Mechanical Engineers, England; Coopers & Lybrand Deloitte, London and Services Ltd., Nottingham, England for Figure 10-1 from their book *Taguchi Methodology with Total Quality* (see Bibliography); the Institute of Electrical and Electronics Engineers, Inc., for permission to reproduce Figure 10-2 from "Computer response time optimization using orthogonal array experiments" by T.W. Pao, M.S. Phadke, and C.S. Sherrerd, AT&T Bell Laboratories, 1985 IEEE International Conference on Communications, June 23-26 (pp. 890-895); and the Society of Automotive Engineers, Inc. (copyright 1984), Figure 9-5.

PART ONE

Tremendous Gains

CHAPTER ONE

The Need for Change

To cut product time-to-market by one-third
To raise quality to world-class levels
To cut the time to ramp up to full production

Never has competition for manufactured goods been keener: Japanese companies have absorbed the increased value of the yen, the newly industrialized countries (NICs) are attempting to raise quality to compete on equal terms with U.S. manufacturers, while the Eastern Europeans are starting to produce a number of cheap products that meet the demands of consumers.

Because of the massive deficits in the federal budget and in the balance of payments, exporting is no longer an option for U.S. manufacturers; it is a necessity. As C. Fred Bergsten, director of the Institute of International Economics and member of the Competitiveness Policy Council, points out: "To cover the large external deficit, the United States must borrow about $10 billion monthly in new money from the rest of the world — and avoid any net withdrawals from the $1.5 trillion stock of liquid foreign assets already in America."

That is a tall order, and Bergsten argues that the only long-term solution is for the United States to export more, and that it is already on the road to doing so. If the deficit is to be turned around, not only must more corporations export more products, but others must produce products that will prove to be import substitutions — the declared aim of General Motor's new Saturn automobile.

Progress, but Not Enough

It is true that progress is being made, with the decline in the value of the dollar an important factor. U.S. manufacturers have increased their share of the world's export markets by about 0.6 percent annually between 1987 and 1990. Some industries are doing better than others in exporting or taking sales back from importers. For example, the U.S. semiconductor makers increased their share of the world's market from 34.9 percent in 1989 to 36.5 percent in 1990. They have also increased their share of the Japanese market from around 3 percent in 1986 to 13 percent in 1990.

Although that is remarkable progress, the increased share of a market that has grown strongly came as a result of U.S. political pressure. In 1986, the Japanese were persuaded to agree to buy 20 percent of all their semiconductors requirements in the United States by 1990. To pressure the Japanese to do so, U.S. authorities imposed punitive sanctions on imports of electronic products made with Japanese semiconductors that amount to around $160 million a year. So it is hardly surprising that the Japanese made efforts to buy American semiconductors.

Other industries have been less successful in turning back the tide of imports and in increasing exports. The U.S. automobile industry is one example: in 1990 General Motors (GM), Ford, and Chrysler posted one-year losses in sales of 3.7 percent, 11 percent, and 15 percent respectively. By contrast, Toyota's sales in the United States were up 15 percent, and those of Honda up 9 percent.

It was encouraging that in 1990, imports of American cars into Japan increased by 50 percent, from 19,084 to 28,602 units. However, 7,700 of the extra units came from Japanese factories in the United States, with the indigenous U.S. makers increasing sales by only 1,794 units, or 12 percent. Honda alone increased sales of its imported Accord coupe by 2,837 units to 7,534, so it is clear that the U.S. makers have a long way to go in catching up with the Japanese.

Lower Dollar Not Enough

Some argue that the fall in the value of the dollar is sufficient to help the United States increase exports and reduce the level of imports. The success of the Japanese in maintaining or increasing market share while the yen has appreciated indicate that a cheaper dollar has had little effect. In any case, there are many more successful producers in the world now than in the early 1980s. Korea, Taiwan, Singapore, and Malaysia are just some of the countries that have improved the quality of their products and that are exporting strongly. Other NICs, such as Thailand, are moving up the learning curve in producing goods suitable for export.

Tremendous efforts are needed to combat this flood of imports, and to increase exporting, but the answer is not simply to repackage the products sold in the United States. When NTT, the Japanese telecommunications monopoly, first opened its doors to American-made products, its executives complained that large corporations submitted quotations for products that did not meet the Japanese specifications — they were identical to the American products. Later, NTT did purchase American digital switching equipment, and American companies started to export telephone handsets and other telecommunications equipment successfully. Even so, some customers complained that the office telephones supplied by American companies were harder to use than the Japanese telephones they replaced.

Better Products the Answer

Clearly, what is needed (apart from the more competitive dollar) is better products that meet the requirements of customers.

To meet these goals, U.S. manufacturers not only need to improve their performance continuously but to ensure they can make better products with shorter lead times and with improved inherent quality. It is no longer merely a question of cutting manufacturing costs — itself a major challenge — but of refocusing the direction of the business so that it responds to the needs of customers. Because of the fierce competition customers will become kings, not tolerating even minor faults; they will be able to find better products at lower prices from other sources. The aim must be to put customers' requirements first. Such a change demands a major shift in corporate culture.

This shift in culture must come from a clear recognition of the tasks and objectives that lie ahead and that must be met. But generating and then managing the cultural change is not easy. It requires confidence that only a wide perspective can support. It demands vision on the part of those who are in command of change; and the clear ability to give priority to the achievement of world-class performance, sensitivity to customers' needs, design and manufacturing leadership, and business competence.

It is no longer sufficient to be classed as "competent." Even excellence in one sector of the business in not adequate, as the trials of Citicorp, Disney, IBM, and Kodak demonstrate. The stakes are so high and the perils of failing so great that all-around industrial excellence is the only goal worth seeking. Inevitably, some organic restructuring cannot be avoided — indeed, it will be welcomed by those who can recognize the rewards in sight.

Work by a number of business management consultants has shown that corporations that have one excellent feature tend to concentrate on that until it becomes sanctified as the main aim of the organization. The result is inflexibility and a failure to detect weaknesses in other areas or any problems that impede the corporation's ability to respond to the market.

To prevent itself from becoming addicted to the keys to past success, an organization must remain open to new ideas and maintain a dynamic structure that continuously responds to the outside world. Open-minded management, flexibility, and better response to the marketplace are among the qualities that make up such an organization.

Chief executives striving to make their company grow and keep up with the competition will find that concurrent engineering (CE) fits in with the characteristics that make for a successful company. It will foster the development of a new structure; it will breed more resourceful, profit-oriented managers in all departments; and it will motivate managers to question where the corporation is going — a prerequisite for success in the nineties.

Japanese Lead

It is clear that Japanese companies are able to develop products much more quickly than their American competitors, and with higher quality. In many industries, their products are the benchmarks. This competitiveness results from a number of factors, such as a loyal work force, dedication to the customers' expectations, and the use of improved methods to develop products. The scale of the problem is shown in Figure 1-1, a comparison of development lead times for similar projects in Japan and the United States.

In this study, which measured six stages of development, the most important differences were in the timing of advanced engineering and process, or production, engineering. The Japanese start advanced engineering just one month after work on the concept has started; in the United States, there is a delay of six months as the concept is firmed up. In Japan, process engineering starts just two months after product design itself; in the United States there is a delay of nine months, again to allow the design to be finalized. Then, in Japan the pilot run is not just

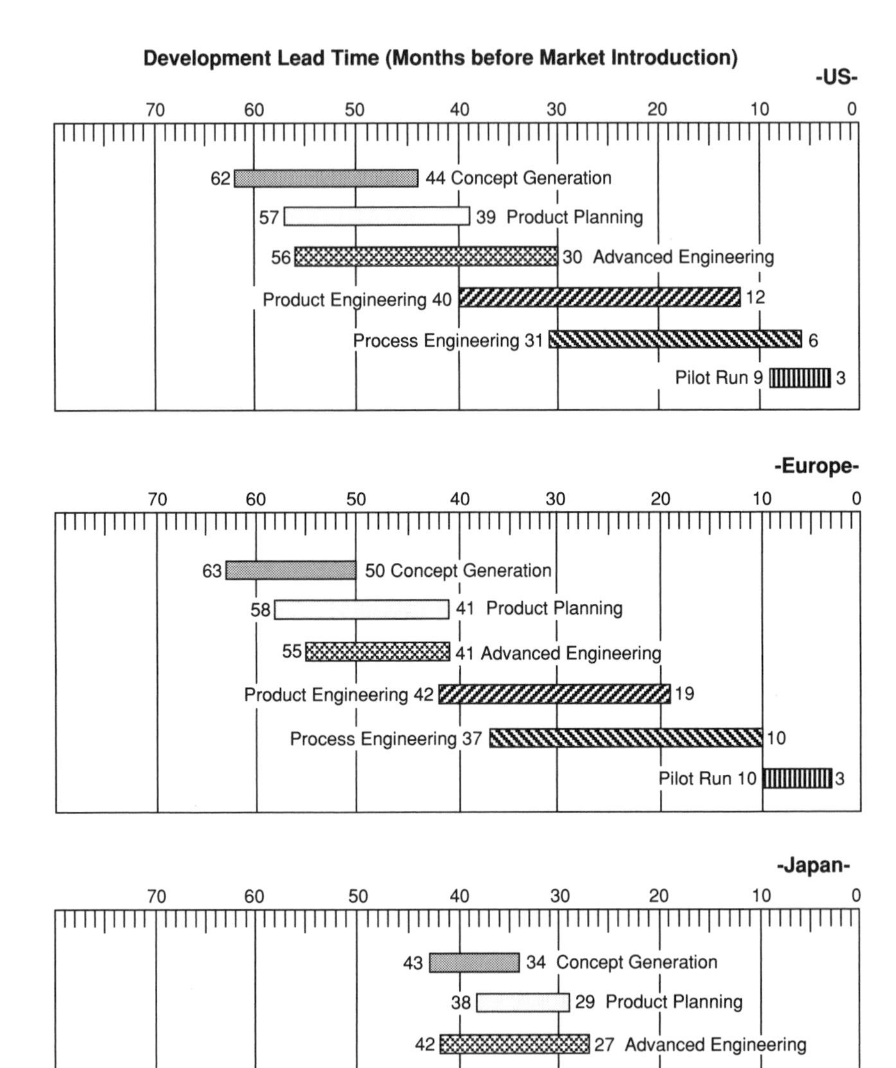

Source: Kim B. Clark and Takahiro Fujimoto, *Product Development Performance: Strategy, Organization, and Management in the World Auto Industry.* Boston: Harvard Business School Press, 1991, p. 78. Reprinted by permission.

Figure 1-1. Because of their team approach, the Japanese can develop products much more quickly than American corporations.

shorter but starts later than in the United States. In addition, most stages are completed more quickly.

Crucial Differences

The differences are crucial. First, because advanced engineering starts so early, the Japanese engineers are aware of every little change made at the concept stage and can thus incorporate them into their thinking early on. At the same time, the process engineers would not be able to start their work unless they were aware of every change made to the design as it was being made — otherwise, they would be designing machines and their vendors building machines suitable for obsolete designs. To be able to respond quickly to these changes, the engineers need to react flexibly and with a sense of cooperation. All these factors result from the adoption of a task force system of project management. Finally, the pilot run starts later because the Japanese have a more completely specified product, with fewer changes being made late in the project. They can therefore afford to wait until the manufacturing equipment is virtually complete.

To close the gap, American companies must emulate their Japanese counterparts by reducing the time taken to develop new products, achieving equivalent levels of quality and similar production volumes from Job One. In the automobile industry, the Japanese are able to take a program from management approval to production in 36 months, against 48 to 60 months in the United States. Their level of defects is approximately one-half that of the United States, and they ramp up to full production within weeks of Job One. The questions facing American managers are: How do we compete? How do we move one jump ahead?

Piecemeal Progress

During the past decade, manufacturing companies have made progress in streamlining operations. Deadwood has been cut out — for example, the Big Three automakers have adopted just-in-time concepts to cut the value of their inventory by $1 billion — and productivity has been improved. Unfortunately, all too often this has been on a piecemeal basis. For example, over a three-year period, a manufacturer might introduce one new product to fill a sector of the market. Because of production problems, volume was probably about 50 percent of that planned for the first 12 months, just when demand was greatest.

Changes made to the design immediately before and during the first few weeks of production reduced quality and increased cost while slowing down production. Meanwhile, the older models, which are less in demand, continue in production in a smooth, trouble-free fashion. However, because of the combination of older plant and lower volume, overall costs of these products are higher than in the new plant.

In another area the same company added some automation, but because of changes in specification at a late stage, the equipment started operating nine months later than expected, and volume ramped up slowly. However, the management team congratulated itself on the elimination of a high level of defects in another plant, a successful program to cut overheads in handling, and the reduction of inventory in the goods receiving warehouse by 27 percent.

With improvements made on many fronts, most on the management team are pleased with progress. But one person is unhappy — the vice president responsible for finance. He or she complains, rightly, that the new product has not yet broken even and that, in the first year, the new plant is not giving the return on investment that manufacturing engineers had claimed it would — so manufacturing costs are still too high. As a result of this criticism, upgrading the efficiency of the new plant becomes a priority, and so the company continues to fight fires instead of planning future products and production in detail. Overall efficiency on a corporate basis has hardly

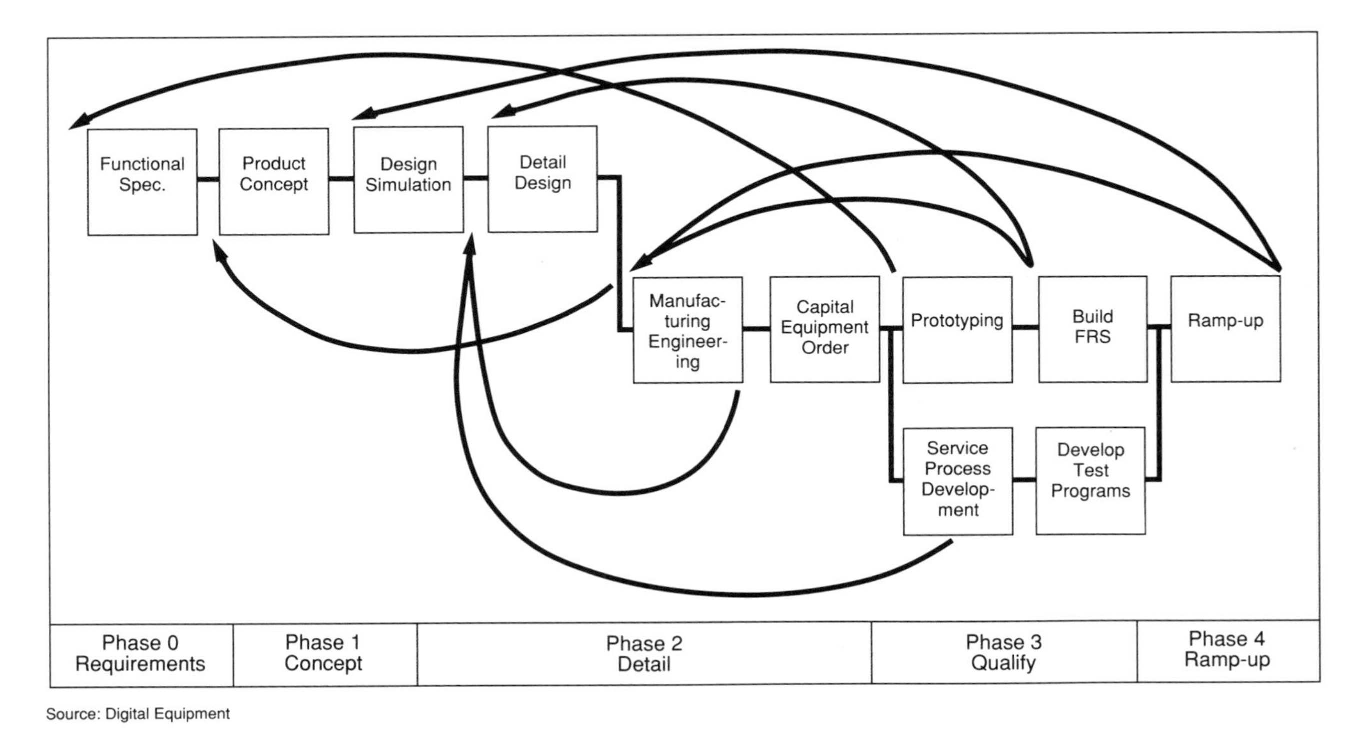

Source: Digital Equipment

Figure 1-2. Backtracking with over-the-fence engineering leads to considerable reworking, which is costly and wastes time.

improved. Such situations are not uncommon in well-managed companies; in other organizations, performance is far worse.

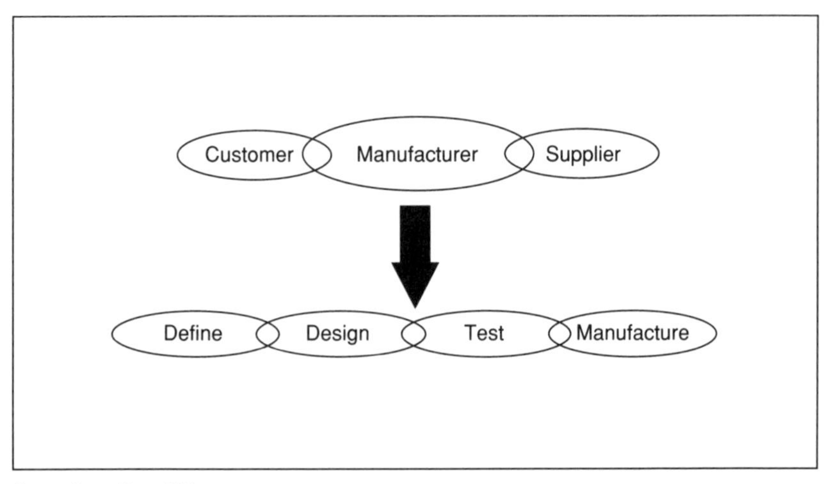

Source: Dowty Group PLC

Figure 1-3. Integration of the various groups and activities leads to improved products.

Comprehensive Approach

Competition is now so keen that it is essential to return to the roots of the business; that demands an improvement of every aspect of the product, from the beginning of design to the end of its service life. Without the product there is no business, and unless the right product is available when the market wants it, profitability will be poor. Therefore, any attempt to sharpen the company's performance must start with the product; cost-cutting, asset-stripping, and improved manufacturing efficiency will be worthless unless the product satisfies the customers' needs, is available when they want it, and has "world-class" reliability and marketability. Moreover, 60 to 80 percent of the eventual cost of the product is committed at the design stage — so it is clear that more funds must be invested at this time to improve results.

Management needs to be assured that each new product will

- be the product customers want at the price they are prepared to pay
- reach the market on time without exceeding the budget — and in 25 to 33 percent less time than at present
- be designed with the highest level of quality and reliability from the outset
- be easy to manufacture in high volume from Job One on machinery that is flexible enough to cope with possible changes
- contain the smallest number of parts and be designed for ease of assembly
- reach sufficient production volume quickly enough to reach a break-even point early

All these targets must be achieved in a cohesive manner, by a team that can balance the trade-offs between these aims against the overall benefit to the project — neither in the defense of the reputation of their department nor with the idea that an unproven method will reduce costs. Now, no assumptions are permitted. A thorough understanding of the market and its niches, the competition, and how to speed up development is a necessity — as are the self-questioning and openness to new ideas that will keep a corporation on the path to a better performance.

CHAPTER TWO

Quickening Change

The solution is concurrent engineering (CE)
CE combines a multidisciplinary task force....
With complete specification at concept
Resulting in short lead time and fewer changes

Concurrent engineering (CE), which relies on a team approach and the adoption of certain specific techniques, is the answer to the problem of improving company performance. Both the team approach and the use of disciplined techniques are essential; neither will provide the potential gains without the other. In addition, records of changes to design, rig testing, experiments, and processes need to be kept meticulously.

Not all exponents of the techniques call what they do *concurrent engineering*. Some talk of *parallel engineering*, and *simultaneous engineering* is used in the automotive industry. The Department of Defense has adopted the term *concurrent engineering* for its CALS (Computer-aided Acquisition and Logistics Support) program for military equipment. Concurrent engineering is the term that most U.S. companies now use, and is the one we will use, because it seems to have gained greatest acceptance in industry.

Corporations such as Honda Motor and Xerox have developed their own names for techniques that are essentially CE. Many Japanese companies do not use CE as such but have been using the basic elements with extraordinary success for 30 years, while others have been using task forces and CE for about 10 years. Nissan took up formal CE only 4 years ago but says it had been using the system "unconsciously" for 30 years beforehand.

Task Force Essential

Whatever the technique is called, the common feature is that each new project is handled by a full-time multidisciplinary task force. Neither a task force from product design nor a committee drawn from different departments that meets from time to time is CE. For CE, the task force normally consists of

- product design engineers
- manufacturing engineers
- marketing personnel
- purchasing
- finance
- principal vendors of manufacturing equipment and components

The task force is permanent in the sense that it normally remains in force throughout the duration of the project, perhaps with some members dropping out to be replaced by others as the stages progress, and in the sense that the members work full-time in the task force. If members spend only Fridays in the task force and the rest of the week working back in their own departments, their work for the task force will not receive the priority it requires.

From the outset, when the design is no more than an artist's sketch, manufacturing engineers in the task force have as much information on the product as anyone else in the team. They can begin planning the manufacturing facilities in the same conceptual way that the product designers are planning the object to be produced — they are working simultaneously. They can interrelate with other members of the team, making

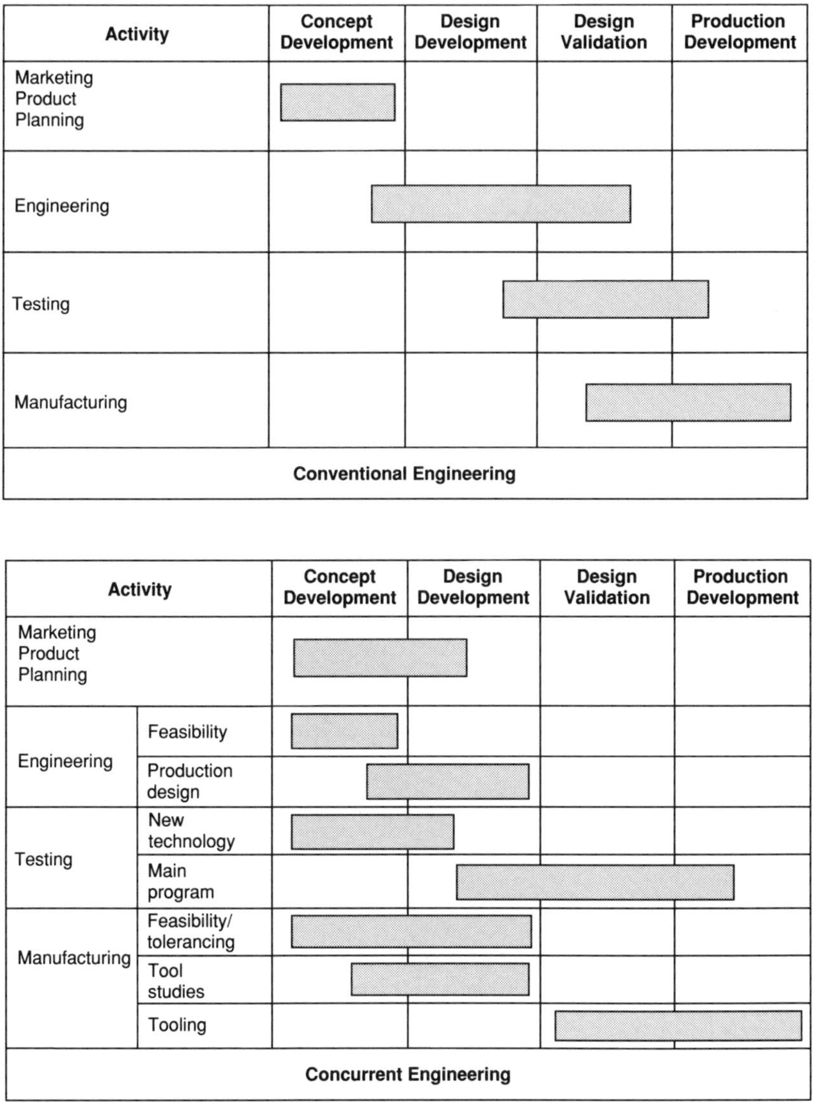

Activity	Concept Development	Design Development	Design Validation	Production Development
Marketing Product Planning				
Engineering				
Testing				
Manufacturing				
Conventional Engineering				

Activity		Concept Development	Design Development	Design Validation	Production Development
Marketing Product Planning					
Engineering	Feasibility				
	Production design				
Testing	New technology				
	Main program				
Manufacturing	Feasibility/ tolerancing				
	Tool studies				
	Tooling				
Concurrent Engineering					

Source: Hawtal Whiting, Inc.

Figure 2-1. With conventional engineering, functions are sequential, but with CE, jobs are done concurrently.

recommendations to reduce cost and the parts count and to raise quality.

The presence of marketing personnel on the team ensures that sales targets are realistic rather than plucked from the sky. Marketing is important because customers' expectations are given far greater weight in CE than with conventional engineering. Marketing will also identify opportunities to make a derivative at a sufficiently early stage for it to be produced at little extra cost.

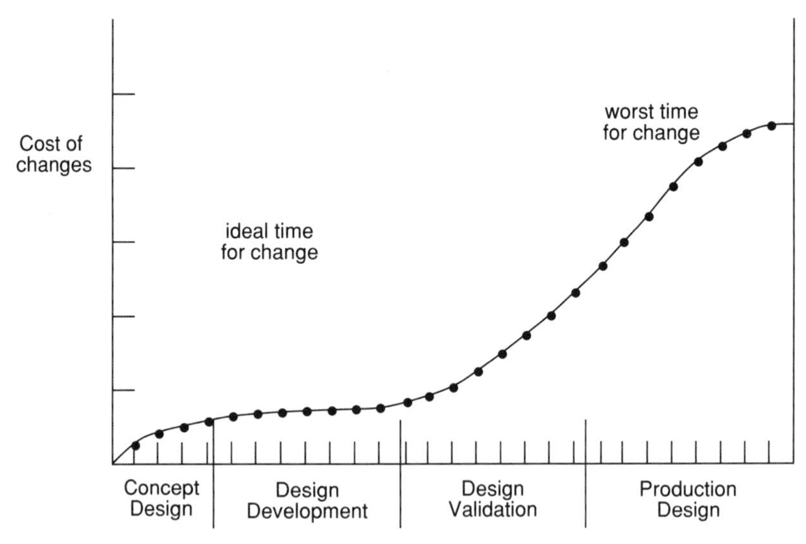

Source: Hawtal Whiting, Inc.

Figure 2-2. Normally, most changes are made at the worst possible time — near the end of the project.

Defining the Product

This approach demands that more time is spent in defining the product than at present; planning too is much more thorough in the early stages. In this way the majority of modifications are made at the design stage, well before prototype or preproduction samples are produced. Therefore, making changes is a routine affair.

It might be expected that this would result in more time being needed to design the product. Far from it; it is in the later stages of conventional projects that most time is spent, as failures in prototypes, changes in engineers' thinking, or revised market projections make it necessary to redesign components. The result is delayed tooling or tools that have to be remade, and the time taken for these changes far exceeds that spent on design definition in the early stages of the project. Where the definition of the product is in great detail, a substantial amount of time is saved.

An important principle of CE is that quality is built into the design from the start, with any features that will be adversely affected by variations in production being designed out. Like total quality control (TQC), CE requires a culture in which everyone is responsible for quality; indeed, TQC and CE fit together like lock and key. Together they open the door to more sales, improved customer satisfaction, and lower overall costs.

Among the changes made to introduce CE, some corporations have closed their quality control departments, transferring the specialists into other departments. This helps demonstrate that there is no passing the quality buck; everyone is responsible. Statistical process control (SPC) still has an important role in the plant, but it is far more important to produce an assembly that cannot be put together incorrectly, and the components of which are inherently durable. A product with fewer parts is likely to be more reliable because there are fewer things to go wrong. When a product fails, it is easier to repair because it can be dismantled and reassembled more easily.

Over-the-Fence Engineering

This contrasts sharply with conventional engineering, often referred to as "over-the-fence" or "over-the-wall" engineering because product design devises a product and then tosses it "over the fence" to the production engineering department. By the time the production engineers are able to suggest radical improvements, the design has proceeded too far for the ideas to be incorporated; in any case, product design resents being asked to redesign a component by an "outsider."

Later, production engineering tosses the product into the factory, and the same problems occur with some aspects of production and quality control. The specialists are brought in only when the design is frozen, and the product design and manufacturing engineers know where the quality problems are but have not eliminated them. Finally, the marketing people are called in. They have already heard the complaints of the production engineers, plant management and quality control department about the product, but they have to muster up enthusiasm and find new ways of selling the product. As if that were not bad enough, they can see many ways in which the product could sell more easily but, once again, it is too late to make changes.

Changes Earlier in Project

Frequently, with over-the-fence engineering, only minor changes are made in the early stages. At this stage design engineers are confident their design is correct for the market, and so they make only a few minor changes to suit manufacturing or marketing. Later, as rig tests by the development department show that further redesign is necessary, the scale of the changes increases. By the time production approaches, product engineering is much more ready to make changes recommended by manufacturing because pilot production shows that rejects are excessive. With this natural progression, it is inevitable that just before the product goes into production some major changes are made somewhere — either to improve the product or to allow it to be produced with a reasonably high yield.

A panic ensues as components destined for early production are reworked so that production can start. The result is products of dubious quality. The people involved in managing these changes spend considerable time and effort ensuring that the changes go through, while the operators in the factory are not sure what the final product will be like until several months after production has started.

The disruptions caused by such an approach are enormous, yet by contrast, even with only three years' experience in CE, some users have been able to move 50 percent of their changes back from the time after the first prototype is made to the stage before a prototype is produced. This is a significant advance in both cutting costs and improving quality. It should be a target for all new users of CE and alone would be justification for its use, but there are many more advantages.

Some managers might see CE as just another buzzword, like flexible manufacturing systems (FMS), computer integrated manufacturing (CIM), or logistics. Heaven knows, there have been enough of them, many thrust onto engineering companies by academics or consultants as the solution to all their ills. CE is not like those at all. It is a way of life — a concept that can be tailored to suit the culture of any organization, however large or small or however structured. Managers can mold CE to suit the way they want to operate and can bring in other techniques that improve efficiency. Moreover, the basic approach and the techniques have been developed by leading manufacturing companies, then honed in the process of increasing market share with better products brought to market in a timely fashion.

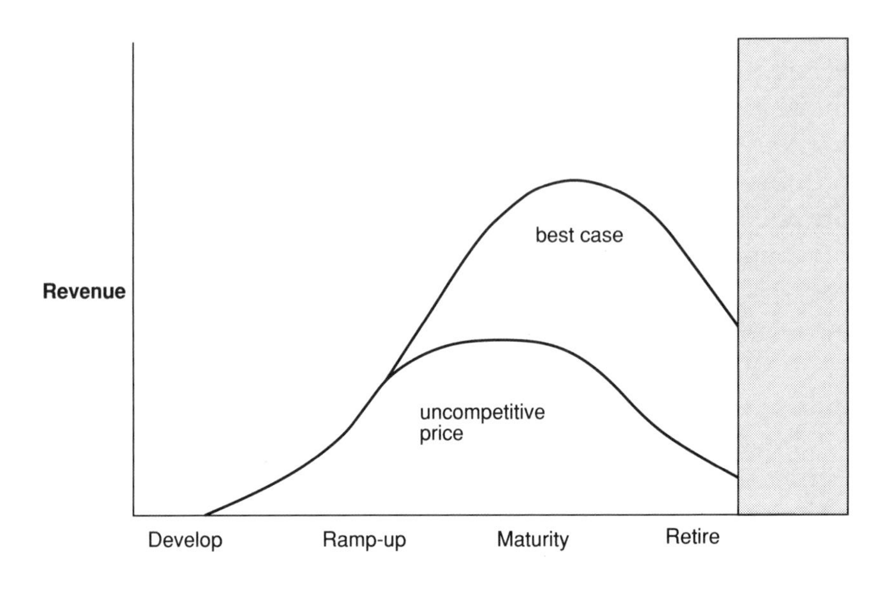

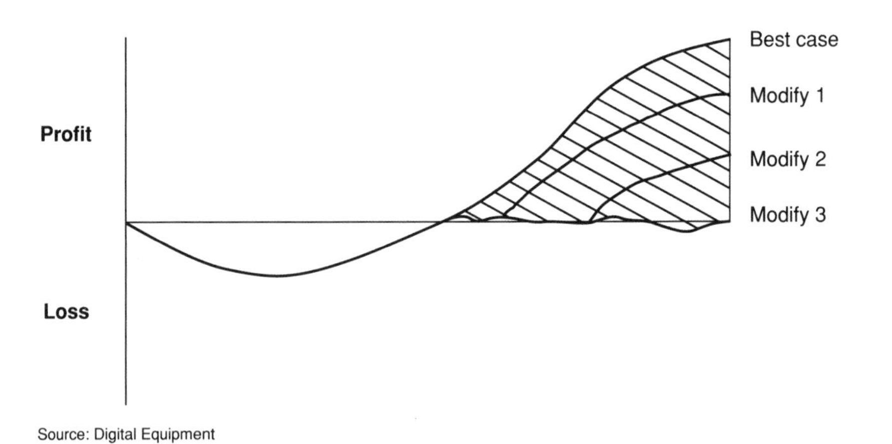

Source: Digital Equipment

Figure 2-3. Uncompetitive pricing results in modifications in production, pushing profitability into the distance.

Wrapping Up All the Trends

CE combines all the trends into a cohesive package and replaces piecemeal improvements with gains in all aspects of the product — and it begins with matching the product to the marketplace. A brilliant product that is too late to market or that does not match the market's requirements is no product at all, merely a waste of effort; this is a truism, but one that development engineers or managers know only too well. Examples of products that never reached the market lie in the recesses of their development shops and in the records.

CAD/CAM, FMS, CIM, JIT, logistics, and concepts to be thought up in the future can be involved in CE projects but with one major difference: from now on, each of these will be given the weight it deserves; it will not be adopted at the expense of other important factors because it is the current buzzword. CAD, for example, is an important adjunct to CE because it allows much simulation to be carried out in parallel.

CE Fits into Company Culture

Some companies may adopt CE with the specific aim of improving design for manufacture and assembly (DFMA). Others may use it to strengthen design by improving the data base of knowledge among their design staff to compensate for a high turnover in staff, or to break down communication barriers between departments. Other managers may see it as a means of obtaining the optimum performance from their vendors as they move from a policy of buying from two or three vendors, who are played off against each other, to one of selecting the preferred vendor, who is awarded a long-term contract.

Although these approaches do not exploit the full benefits of CE from the outset, they will lead to major changes in the way a company operates and to broader use of CE, which can fit into any company — large or small, highly structured or loosely structured, national or multinational. However, the use of CE will change the way in which a company does business, and management needs to be ready to let it develop into its new structure.

Rigid adherence to a hierarchy could put strains on the task forces that negate some of the advantages. Indeed, some large corporations have had difficulties in implementing CE successfully because they had adopted a highly efficient but rigid management structure and were reluctant to see that efficiency eroded. They did not have sufficient imagination to see that the potential benefits outweighed the risks. They were no longer operating their corporation as a learning organization but had become ensnared by their success in creating an efficient corporate structure.

Tremendous Gains

CE is a relatively new approach to project management, but already impressive gains have been reported. Northrop, which produces military equipment, has developed CE partly as a result of its involvement in the CALS initiative and claims substantial gains for CE in the production of its fighter aircraft. Specifically, CE has reduced:

- the number of modifications generated in-house by 75 percent
- the amount of work-in-process by 50 percent
- the production process time for each fighter plane by 33 percent

Digital Equipment reduced the overall time to take a new product from concept stage to full-volume production from 30 to 18 months, and it is now heading for 12 months. It also brought forward the break-even point by some 12 months — an

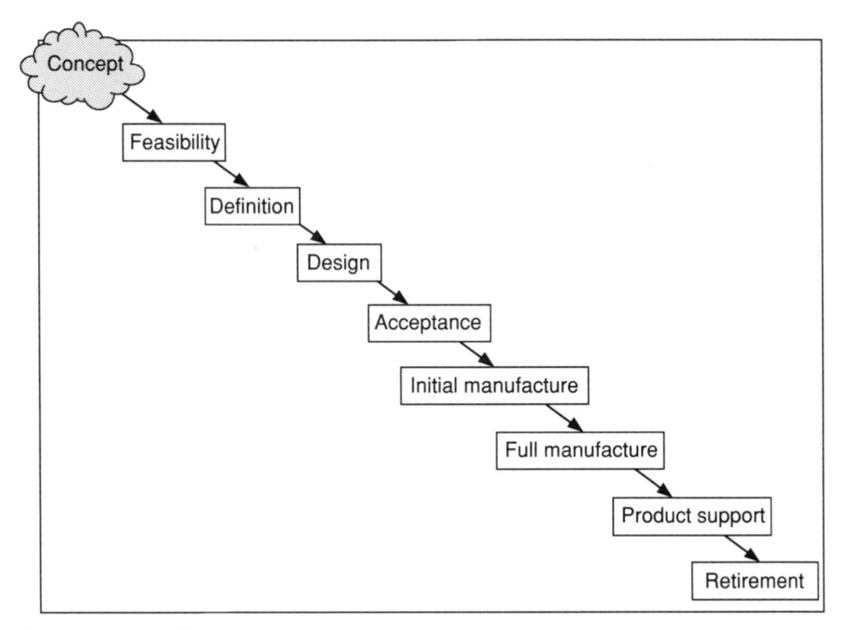

Source: Dowty Group PLC

Figure 2-4. CE cuts costs but alters the timing of expenditure; more is required at concept to definition, less at initial manufacture, and more in product support.

extremely valuable achievement in an era when product lives are shortening every month. Digital was so concerned about the need to reduce the time each new product was taking to break even that it defined the lead time for a new project as the period from board approval to the point at which full-volume production is reached — not the start of production or sales, as is more common.

In the development of a new minicomputer the following gains were reported:

- Time from concept design to volume manufacture was reduced by 60 percent

- Cost over life of product was reduced by $75 million
- Assembly time of magnetic tape drive was reduced by 55 percent
- Parts count of magnetic drive unit was reduced by 52 percent

Another company has cut the typical lead time for a new product from six to four years. These gains demonstrate some of the benefits of CE. It is true that some of the improvements to the products could theoretically have been made by the adoption of techniques such as design for manufacture and assembly (DFMA). In practice, however, with the existing structures in manufacturing companies, such gains are unlikely to be achieved without CE simply because the rigidities of the system prevent the combined effort of personnel from being exploited. For example, by the time the production engineers are ready to introduce a new process, it is too late to implement it at the planned start-up of production, with the result that the product will be late onto the market, and the budget exceeded.

Multidisciplinary Approach

With the task force approach that is fundamental to CE, the whole is far greater than the sum of the elements. That is why CE will cut time-to-market and the cost of programs while bringing improvements in product design, quality, ease of manufacture, and performance in the field. However, CE is not just project management by task force under another name. Vital elements include

- multidisciplinary task force
- product defined in customer's terms, then translated into engineering terms in considerable detail
- parameter design to ensure optimization for quality
- design for manufacture and assembly (DFMA)

- simultaneous development of the product, the manufacturing equipment and processes, quality control, and marketing

Who should adopt CE? Virtually any organization involved in manufacture on any scale; the small firms with 10 to 50 employees are probably doing it unconsciously, and if they are not doing it, they should be. To most companies involved in manufacture, the decision of whether to use CE is theirs. To those supplying weapon systems to the Department of Defense, there is almost an obligation to use CE, which is an integral part of the CALS initiative covered in Chapter 4.

Summing up, CE combines a task force approach to project management with a number of specialized techniques that ensure the design is optimized — from all aspects, not just function. These include manufacture, assembly, maintenance/servicing, spares, and lifetime performance, including recycling. If a change in legislation or in the market adds new factors — as with legislation on recycling or energy use — a specialist can be added to the team. Only with CE can product development be flexible enough to adapt to any changes.

To gain the benefits of CE, companies must apply the concepts with unswerving enthusiasm; management must allow the task force to manage its product, and must trust and involve vendors where necessary. CE is no "quick fix" to superimpose on an inefficient operation; it is a tool with which to root out inefficiency and to gain the utmost from the skills in the organization. But it requires imaginative managers if it is to become an engine of growth.

CHAPTER THREE

The Key to Japanese Success

Team approach part of the culture
Quality function deployment (QFD) and
Taguchi's techniques developed in Japan
New model lead times cut to less than 36 months

Over the past decade, the Japanese have not only made big gains in their share of markets, but they have impressed with their ability to introduce models with high-technology specifications. In addition, they have shown an ability to match customers' needs closely. Why? Largely because they use concepts that foreshadow CE.

As mentioned in Chapter 1, a comparison of 6 American projects and 12 Japanese projects to develop new products showed that the Japanese required 19 months — 30 percent — less lead time than their American counterparts. The Japanese are engineering in parallel, and are reaping the rewards. For example, Komatsu combines the use of the team approach with QFD, which is discussed in Chapter 7, and as a result has cut lead time dramatically. In one case, it developed four ancillaries for its dump trucks in 12 months, whereas the industry norm was three to five years.

Not all leading Japanese companies operate a formal task force system, and many do not embrace the idea of CE itself, even though they may use similar techniques. However, their companies are structured in such a way that teamwork is inherent in policy decisions, in new projects, and in day-to-day operations.

Group Culture

Japan's culture is based on the need to find a consensus on a course of action, be it in government, in a company, or in the community. Companies have large boards of directors, and each vice president has a faction of supporters, including some junior directors. All the vice presidents and directors must agree on a new project before it receives the go-ahead. Contrary to the popular view, the consensus is usually reached only after considerable argument in which all the managers involved have their say, which may show that one faction wants to do something diametrically opposite to the wishes of another faction.

Eventually, after public debate and private discussions, the consensus emerges. After that, everyone is committed and will give their all to the project. There is no political back-stabbing, as one manager waits for another's idea to fail so that he or she can gain promotion. Of course, there are internal political battles, but loyalty to the corporation and to the project takes priority.

Another important aspect of Japanese companies is their thirst for knowledge about the markets they serve. Japanese companies are true "engines of inquiry" at all levels, and this is one reason why they are successful. Japanese corporations are kept well informed of their markets by their people in the field, by the embassies and Japanese trading houses, and by some of the most talented lobbyists in the business. Until very recently, the Japanese considered themselves to be catching up to the West, and so they have been far more willing to listen to the customer than have Westerners. One reason for this is that Japanese corporations do not place people in the departments where their university training would be of direct value. Instead, they rely on the university to educate people so that they are suitable for training. In an overseas sales section, it is not uncommon to find one person who majored in business studies, one in law, another in chemistry, a couple in economics, and one in English.

The people who are collecting data for the company and formulating sales programs are not self-styled experts relying on what they learned about marketing in college. They are far more receptive to new ideas than is generally the case in the United States. In any case, jobs are not defined precisely, and the work going on in different departments overlaps, but not in a competitive way. Therefore, the environment in Japan is conducive to the task force approach, even though many of the meetings may be informal and may not be considered to be a task force in the way we define it for CE.

QFD and Taguchi Born in Japan

Significantly, both quality function deployment (QFD) and the Taguchi methodology, which are discussed in greater detail in Chapters 7 and 9, respectively, were developed in Japan. Genichi Taguchi developed his techniques in the 1950s, while QFD came ten years later. Both were intended to improve the quality of the product and to make design engineering more efficient. Since Taguchi was working at NTT, the Japanese telecommunications monopoly, when he developed his system, it was picked up by the "NTT family" of vendors that includes such names as NEC and Oki Electric, and spread through the electronics industry.

QFD was adopted first by Mitsubishi Heavy Industries in the early 1970s, and was picked up by Toyota Auto Body in 1977. Later, other companies found that it fitted in neatly with their efforts to know the customer's view, and so QFD came into general use. Now, many leading companies have woven QFD so tightly into their method of operation that engineers consider it an in-house development.

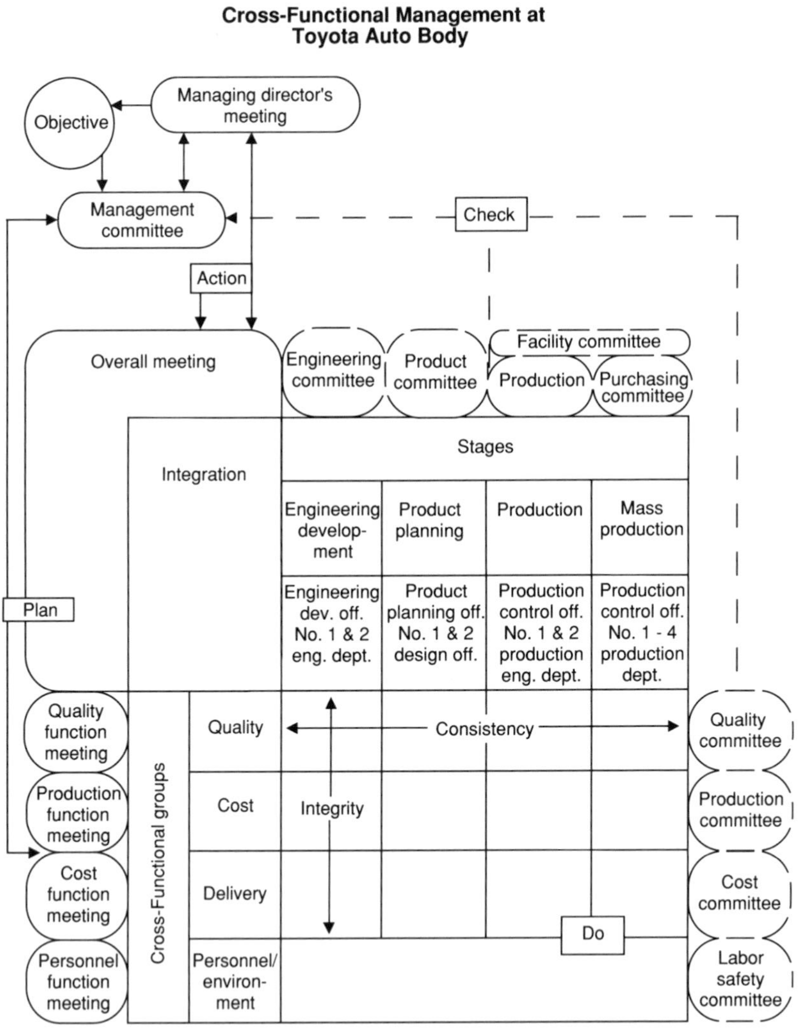

Cross-Functional Management at Toyota Auto Body

Objective
Managing director's meeting
Management committee
Check
Action
Plan
Overall meeting
Engineering committee
Product committee
Facility committee
Production
Purchasing committee
Integration
Stages

Engineering develop-ment	Product planning	Production	Mass production
Engineering dev. off. No. 1 & 2 eng. dept.	Product planning off. No. 1 & 2 design off.	Production control off. No. 1 & 2 production eng. dept.	Production control off. No. 1 - 4 production dept.

Quality function meeting
Production function meeting
Cost function meeting
Personnel function meeting
Cross-Functional groups
Quality
Cost
Delivery
Personnel/ environ-ment
Consistency
Integrity
Do
Quality committee
Production committee
Cost committee
Labor safety committee

Source: A.A. Aswad and J.W. Knight, " Comparative Aspects of QFD and Simultaneous Engineering," in vol. 1 of *Proceedings of the 21st International Symposium on Automotive Technology & Automation (ISATA)*, Wiesbaden, November 1989, p. 152.

Figure 3-1. Toyota Auto Body adopted a matrix approach based on QFD in managing new product development.

The involvement of manufacturing engineers in projects from an early stage is easily achieved in Japan because most of the large manufacturers make their own special-purpose machines, or have subsidiaries in this business. They do not need to rely on vendors, as is the case in the United States. Within the Honda Motor Company group, for example, Honda Engineering makes machine tools, dies, welding fixtures, and some robots; Nissan builds most of its special equipment and robots; Toyota has a subsidiary, Toyoda Machine, and an interest in other machine tool companies; Mazda has a machine tool company among its subsidiaries; and Nippondenso makes its own manufacturing equipment, including robots, in-house. Among the electronics companies, Matsushita Electric, NEC, Oki Electric, and Sony produce much of their own manufacturing equipment, including robots. This situation is the norm rather than the exception among major Japanese manufacturers, so it is easy for the makers of the equipment to be involved in brainstorming sessions to find the best production methods at an early stage in the project.

Overall, Japanese companies control far more of their manufacturing than do their American counterparts. Also, most of the vendors that supply directly to a major company are either subsidiaries or affiliates in some way. "Subsidiary" means not that the major manufacturer holds more than 50 percent of the stock but that it does hold some stock — usually between 5 and 15 percent. It will not have a stake in the smaller vendors, but these companies will nevertheless consider themselves part of the group and will depend on one major customer for 60 to 90 percent of their business.

In Japan, the group vendor may be totally responsible for the design of some components, whereas in the United States the design is usually controlled in-house, even though consultant design houses may be employed. Also, the Japanese rely more on their vendors to do their own testing and to supply proven

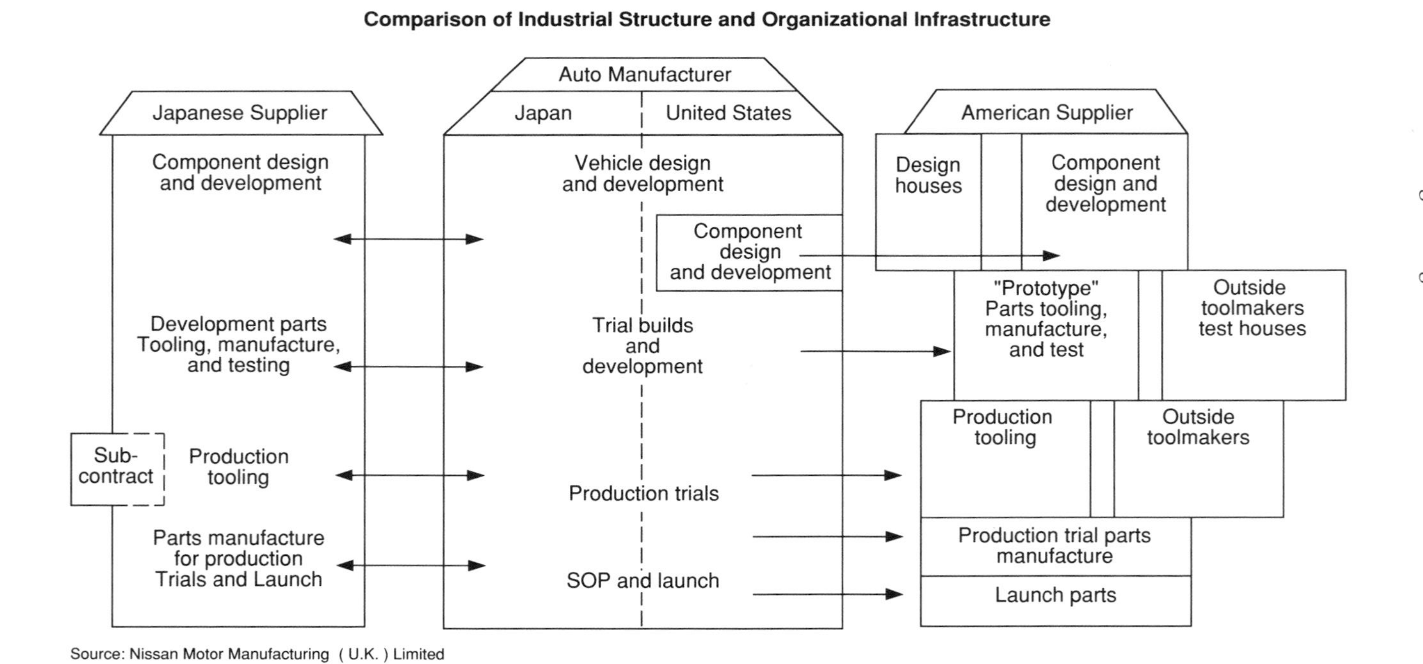

Figure 3-2. There is a marked difference between the structures of Japanese and American auto companies and their vendors.

components. The different structure has an effect on the way that teams are built up. In Japan, it happens naturally, with close liaison between manufacturer and vendor, because both see themselves as part of the same group. Therefore, when Mazda planned the transfer line for its V-6 engine, it bought the broach for the main faces from Ingersoll Milling, but most of the machinery was built or specified by Toyo Kogyo Machine Tool, its subsidiary.

A Team that Does Not Look Like a Team

Neither Toyota nor Mitsubishi Motors operates formal multidisciplinary teams to control projects, preferring the system in which the program is under the control of a chief engineer or project engineer, in much the same way that Western car manufacturers control projects with vehicle engineers working within a matrix of component teams. However, those groups work as teams, and because of the way the companies operate, they form nuclei of bigger task forces that include experts from other departments.

One of the main reasons that Toyota merged Toyota Motor Company and Toyota Motor Sales in the mid-1980s was to shorten lead time and improve the feedback from the customer to the product engineers. In the 1970s, there was considerable friction between the two companies, with Toyota Motor Sales going so far as to postpone the marketing of a front-wheel drive car developed by Toyota Motor Company for a year or so. The decision-making process was cumbersome, with the result that Toyota was not responding to changing fashions as quickly as its smaller and more nimble Japanese competitors.

Once the two companies were merged into Toyota Motor Corporation, the structure was reorganized so that marketing information could be acquired much more quickly. With that background it is not surprising that Toyota insists on new model projects being under the control of product engineering.

There is a team for each new project under the control of a chief engineer, with 10 to 20 people in the team, mainly from product planning. This team works with marketing in the conventional manner, while specialist vendors of components and machines are brought into meetings on a regular basis from an early stage. Meetings between people from different departments are held regularly, and in effect these are small committees from the task force.

Moreover, Toyota has adopted techniques of concurrent engineering in the pure sense, with as many operations as practical being carried out in parallel. To further cut lead times, Toyota has developed a CAD/CAM system with a common data base for design and for the manufacture of press tools and welding fixtures. This is typical Japanese CE — unconscious CE. Nissan has also been using elements of CE for years. For example, it has established *kohren,* or factory liaison boards consisting of vehicle designers and production engineers, which work in teams on new designs for several months before the design is finalized.

Honda's Decade of Expansion

Among the Japanese car makers, Honda, Mazda, and Nissan have adopted CE wholeheartedly in a manner that resembles the Western approach. Honda was one of the first to adopt CE anywhere, with its SED (Sales, Engineering, Development) system of product development. Honda does not call its system concurrent engineering; when the then president Kiyoshi Kawashima introduced the SED system into Honda in the late 1970s, concurrent engineering was not even recognized as a methodology. But that is just what SED is — a product development team combining sales, engineering, and development division members.

Kawashima wanted Honda to be able to respond more quickly to customers' requirements and to competitors' moves,

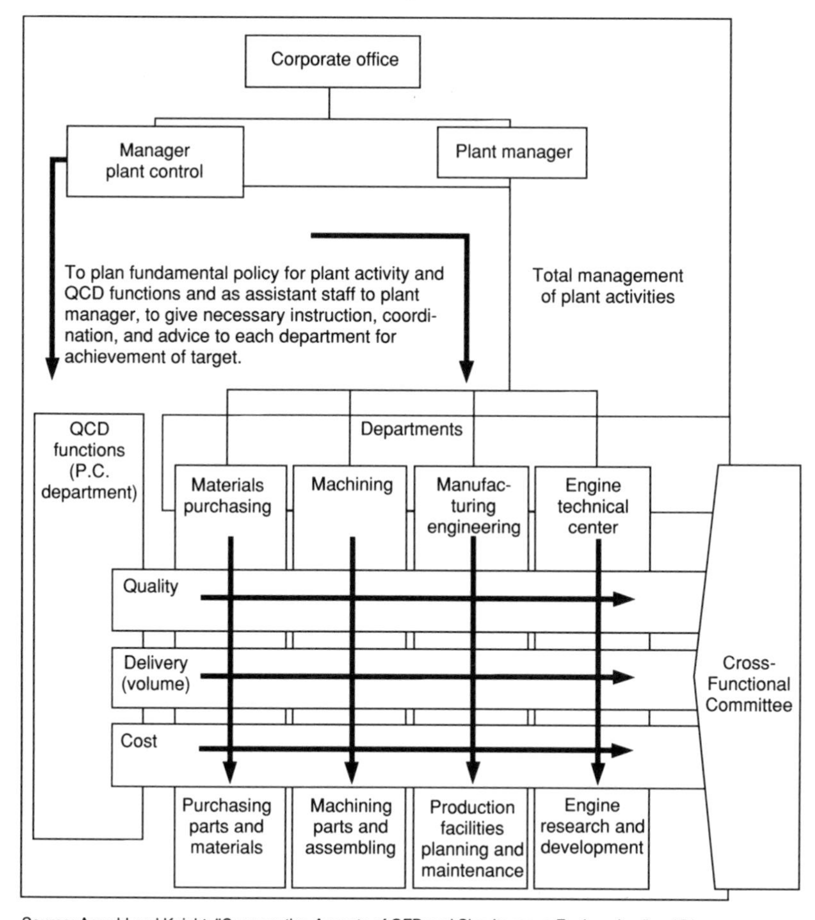

Source: Aswald and Knight, "Comparative Aspects of QFD and Simultaneous Engineering," p. 151.

Figure 3-3. Komatsu has a flat management structure suitable for the team approach.

and sought improved quality. It was decided that a multidisciplinary task force approach was needed. At that time Honda was a relatively small Japanese carmaker, building the Civic,

Prelude, and Accord car ranges and some lightweight trucks. In 1980 it produced 956,900 vehicles, including 107,000 light trucks, all in Japan. It was also having difficulty breaking into the market for automobiles with engines bigger than 1.3 liters in Japan, because the general public saw Honda as a maker of motorcycles, so durability and reliability were not considered important.

Unparalleled Seller in the United States

Ten years later, Honda's reputation as a manufacturer of high-quality automobiles is secure, and output has more than doubled to two million units a year, one-third of which are built outside Japan. Its market share in Japan has increased sharply and sales in the United States have grown to an amazing 855,000 units — equal to its total passenger car production in 1980 and just 6,000 units below Chrysler's sales. The Accord was the best-selling car in the United States in 1989 and 1990, while Honda's U.S. output was over 400,000 units. Moreover, Honda sold about 80,000 more cars in the United States in 1990 than Toyota, despite Toyota's worldwide production of around four million vehicles a year.

As a result of its use of SED, Honda has been able to expand the range of cars significantly, and add many derivatives. It now produces the following ranges of passenger cars:

- Today midget car and Beat midget sports car
- City minicar three-door hatchback
- Civic range of two-door sports coupe, three- and five-door hatchback, and four-door sedan
- Concerto four- and five-door sedans
- Integra three- and five-door hatchbacks and two- and four-door sedans
- Accord four-door sedan, some with transverse four-cylinder and some with in-line five-cylinder engines,

and two-door coupe (the coupe is built in the United States, and some are exported to Japan)

- Prelude two-door sports coupe
- Legend four-door sedan and two-door coupe
- NSX two-door mid-engine super sports coupe, produced at the rate of around five thousand a year

Not only is this a substantial range of vehicles, but in line with the Japanese tradition, most are rebodied at four-year intervals. In addition, during the 1980s, Honda produced new engines for all its existing engine families and introduced two extra engine types — an in-line five-cylinder for some versions of the Accord and the V-6 units for the Legend and NSX. If that was not enough, Honda engineers devised an innovative four-wheel steering system, their own anti-lock braking system — first introduced in the early 1980s — an electronically controlled fuel injection system, and a four-speed automatic transmission, and was able to produce and maintain world-beating Grand Prix engines for most of the decade. The extent of the expansion, and the range of successful technical developments made by Honda without reliance on vendors to develop systems for it, is unparalleled in the period. Were Honda to have used over-the-fence engineering, the work load would have been too great, and some of the developments that helped fuel the expansion would have been impossible.

SED the Key to Success

Honda SED, which has been the key to the corporation's remarkable success in the past decade, symbolizes the three main divisions within Honda:

- Sales, including service
- Engineering, for Honda Engineering, the company that makes manufacturing machinery and tools

- Development, for Honda Research and Development, which is responsible for all new product development

Each division provides members of a team responsible for engine, transmission, body, and chassis, while the Research Division of Honda R&D provides input on new materials and technical developments. For example, the Research Division carried out a study of the viability of an all-aluminum body and the type of fixings required for the NSX supersports car. It also recommended the use of titanium connecting rods for the 3.0 liter V-6 engine in that automobile. As is common in most CE teams, a member of the R&D team is project leader.

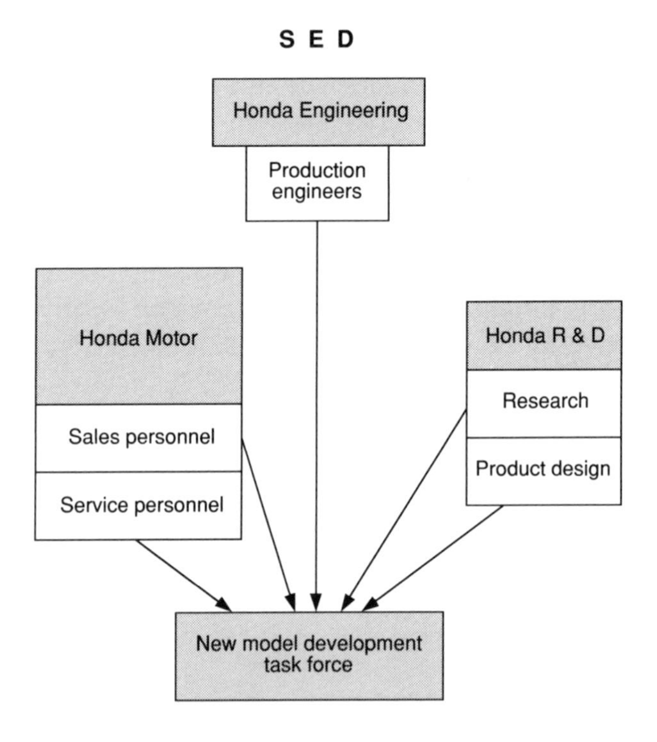

Figure 3-4. In Honda's version of CE, Honda Motor, Honda Engineering, and Honda R & D all send members to each task force.

Actuality, Dismantling

Sales acts as the voice of the customers, and the task force translates the results into specifications following the techniques of QFD. At this stage, Honda emphasizes that the commonplace component is just as important as the exciting new technology. To this end, it has developed two special techniques — *genba genbutsu,* which means actual place, actual article; and *zenbara,* which means dismantle completely.

Genba genbutsu, an idea of Soichiro Honda, the founder of Honda Motor Company, is intended to ensure that the task force knows all there is to know about the product or concept. Thus if the task force is looking at a development in the chassis it will watch how a vehicle behaves on the road, and how ordinary drivers respond to certain situations. For example, when Honda was investigating what type of four-wheel steering to adopt, everyone involved went into a parking lot to observe and discuss how people actually parked their cars. Of course, once a prototype system had been developed, Honda went through the more conventional stages of testing it, with the public as the test drivers, and gauging reactions of professional drivers, including the motoring press.

If the task force is investigating a problem involving a pair of components that are difficult to assemble, or a fault in a component, it will go to the shop floor and see the product in production. Then the members will discuss what they saw. If the components are difficult to assemble, the team might even try to assemble the parts themselves, and they will certainly ask the operator for his or her opinion. Not only is the team more likely to succeed in solving the problem than if the manufacturing engineer goes alone to the shop floor, but their interest will encourage the operator to make some positive contribution, and will raise his or her morale.

Zenbara, on the other hand, was developed specifically for the task force, and is intended to improve knowledge of the product and to inspire new ideas. The assembly being discussed is placed on the table at a meeting and is dismantled piece by piece. Prior to each stage, however, the task force is required to describe the component in the fullest detail, down to the smallest chamfer, and to comment on the usefulness or otherwise of the feature. Because of the need to contribute, members are likely to query features they would normally accept as being necessary. The boredom of some at studying some aspects of the component also brings out some criticism that can be made constructive. In this way, components are often simplified.

With these two techniques, all members of the team become familiar with the product and what it does, and why every feature is needed. If some features are redundant, that too will become clear. Honda executives say that such a hands-on approach is important since it keeps the more senior members of the staff in touch with the product. By contrast, they find that senior engineers in Western automotive companies spend all their time shuffling paper and no time handling or inspecting the product.

Guest Engineers Improve Design

When it adopted SED, Honda also established the idea of "guest engineers," in which engineers from vendors work at Honda R&D alongside the engineers involved on the same product. For example, a design engineer from the vendor that makes shock absorbers might work alongside the Honda designer who is designing a suspension that requires a new type of shock absorber body. This approach forestalls any "them-and-us" attitude among staff and ensures that the vendor is a true part of the team. Since most of Honda's vendors are Honda group companies, it was relatively easy to institute this system. Engineers from non-group companies such as Nippondenso, the auto-electrical company that is part of the Toyota group, also work as guest engineers at Honda from time to time.

Honda's SED has a formal decision level, which starts with the R&D team. In most minor cases, whether the problem is a fault or a potential improvement, the team will make the decision on its own. When it cannot agree or when the decision involves areas of the product not the responsibility of the subteam, the SED task force meets. If the problem still cannot be resolved, or if the budget, a new product line, or another policy matter are involved, then the decision is made by the board of directors. In practice, however, many decisions are made during informal meetings of the task force.

Once the car or assembly nears production, the membership of the SED task force changes. Before production starts the plant manager will become a team member, bringing along some of his or her specialists. Then, when production starts, the plant manager takes control, and new members will be co-opted to deal with problems that arise in production. Service engineers are involved together with only one member of R&D, the others moving on to new projects.

What progress has Honda made in shortening its new model lead time with the aid of SED? It has reached the stage where the complete project from preconcept to production takes 32 to 36 months, and it intends to reduce this to 24 months in the next few years. That Honda can achieve such a timetable is borne out by the fact that when it worked with the Rover Group on the 800 and 200/400 series, it was able to start up volume production 12 months ahead of Rover, even though the initial concept work and much of the detailed design for Honda and Rover models were done at the same time. In addition, Rover had by that time made a lot of progress in reducing lead time as well — although it started from a lead time of 6 to 8 years.

Mazda's Decade of Experience

Mazda is another company that adopted the task force concept many years ago — in 1978, at about the same time as Honda. At that time, it was recovering from its near bankruptcy in the first energy crisis and was looking at ways to ensure that its future models were in tune with what customers wanted, or might want if conditions changed. In other words, it saw the task force approach as one plank in its rejuvenation program.

This was no surprise, because Mazda's troubles in 1974-75 stemmed from its decision to give priority to the development of the Wankel rotary engine, despite the product's poor gas mileage and unreliability. Demand was so poor that Mazda sent many engineers out to work temporarily in its sales offices and at its dealers. In trying to sell the huge stock of rotary-engined cars, they found that many customers were simply not interested in what type of engine was under the hood. They wanted a compact, economical, and practical car. Mazda realized it needed to listen more carefully to its customers, and so it embarked on the development of the first GLC hatchbacks. Later on, Mazda adopted a flexible approach based on a multidisciplinary task force, with around 15 people being involved at the start of the project. When a particular problem is encountered or a critical stage reached, additional people are brought in for a few months. Usually, the project team remains in charge of the vehicle until production starts.

However, the team leader, who is an engineer from R&D, remains responsible for the model as long as it is in existence, or until he or she is assigned to a new responsibility. Generally, the team leader will remain responsible for the vehicle throughout development and production, and if a new job is to be assigned, the handover takes place when the succeeding model is at the preconcept stage. This is closer to the "profit center" approach of some American corporations, where the task force is responsible for the success of a product from preconcept to retirement.

Thus, Mazda's main thrust is in the use of a task force, although it does use QFD in the development of new models. In some cases, however, it relies on the project planner's knowledge instead of the customer's voice. Mazda has also used Taguchi in a few instances but has yet to adopt it as a matter of company policy for specific purposes. Nor does Mazda involve any vendors of components or machine tools in its task force. To ensure that they remain close to the project, it puts someone from purchasing on the team, and this person works closely with vendors.

With its task force system, Mazda has been able to speed up the development of its new model programs. Overall, the development of a new model from the decision to start the project to production takes about 42 months. The first 18 months are spent taking the initial idea to approval by management of the clay model. An additional 24 months is needed to take the concept to production.

Nissan Takes Up CE with Gusto

In an attempt to overcome the weakness of its product line at the time, Nissan adopted CE as a discipline in January 1987. Nissan had been losing market share in Japan for several years, and had also suffered a massive internal political row over whether it should build a plant in the United Kingdom. CE was adopted as a means of raising the company from its poor position. Since then it has been able to identify a few niche markets and to open them up successfully well ahead of Toyota, partly as a result of CE.

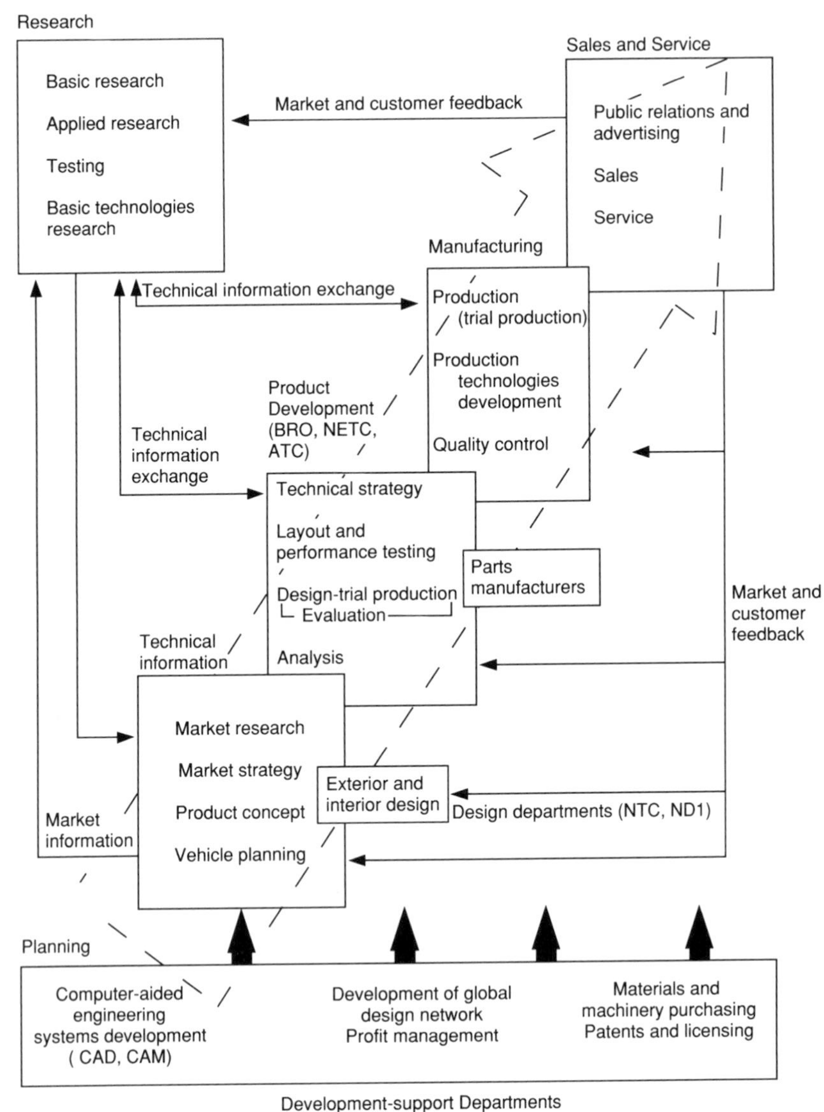

Source: Nissan Motor Company Limited

Figure 3-5. Nissan's version of concurrent engineering is designed so that it can respond more quickly to changes in the market.

In adopting CE, Nissan coined the expression *simulconscious,* or *conscious CE,* to distinguish it from the unconscious CE the company had been using for 30 years. A products and market strategy office was set up to determine demand trends. It is responsible for planning the company's product lines, power trains, and marketing. There is a product planning and marketing group, with one task force for each model. The task force includes members from production, quality control, and testing. In addition, at the beginning of each project and at times during its progress, other specialists join in meetings. Like Toyota, Nissan places control firmly in the hands of product engineering, but unlike Toyota, it includes members from other departments in all projects.

Now that Nissan has established CE in Japan, it is extending the technique to its operations outside Japan and will adopt the approach at its overseas engineering centers. Indeed, the company considers CE essential because the structure of industry in the West differs from that of Japan.

It has therefore started a program to explain to vendors how it develops products and how it wants vendors to become involved. It is requesting major vendors to increase the amount of design, development, and testing carried out in-house. In other words, it is attempting to persuade them to operate more like Japanese vendors. It has also adopted CE for projects outside Japan with multifunctional teams in control from the start. Significantly, it wants its vendors to be more flexible than they are at present, by creating an environment capable of managing frequent change. This of course is one of the essential features of any successful management approach.

To foster the development of the best of new manufacturing and product technologies for use in all its vehicles, Nissan also established a concurrent engineering center in 1988. Among its personnel, 30 percent were drawn from product engineering; 30

percent from the central research laboratory; 15 percent from the styling department; 15 percent from the assembly plants; and 10 percent from the test and prototype sections. The teams in the CE center are working on projects such as

- structural adhesives for bodies
- precision castings
- laser technology
- the use of plastics for structural components such as vehicle bodies and housings
- modular vehicle assembly

This is a new approach, since the work is applied R&D, an area where production engineers are rarely involved, and if they are, their role is that of adviser rather than full team member. At Nissan, all members of the teams work as equals, so the breakthroughs required in these new technologies are more likely to be achieved here than in corporations that carry out the work behind closed doors in an R&D department staffed entirely by product-design-oriented personnel.

Electronics Manufacturers

Matsushita Electric Industrial, the huge company behind the Panasonic name, is at a relatively early stage with conscious CE, but has been using a number of elements for some time. Sales agents have direct contact with the design department, so that the voice of the customers is strong. Like many Japanese corporations, Matsushita has concentrated on reducing lead times in production.

There is considerable overlap in the periods when product design and engineering of the production system are carried out. In the case of a project to build industrial controllers, the production engineers started designing their equipment after the product designers had only been working on the project for 2 months

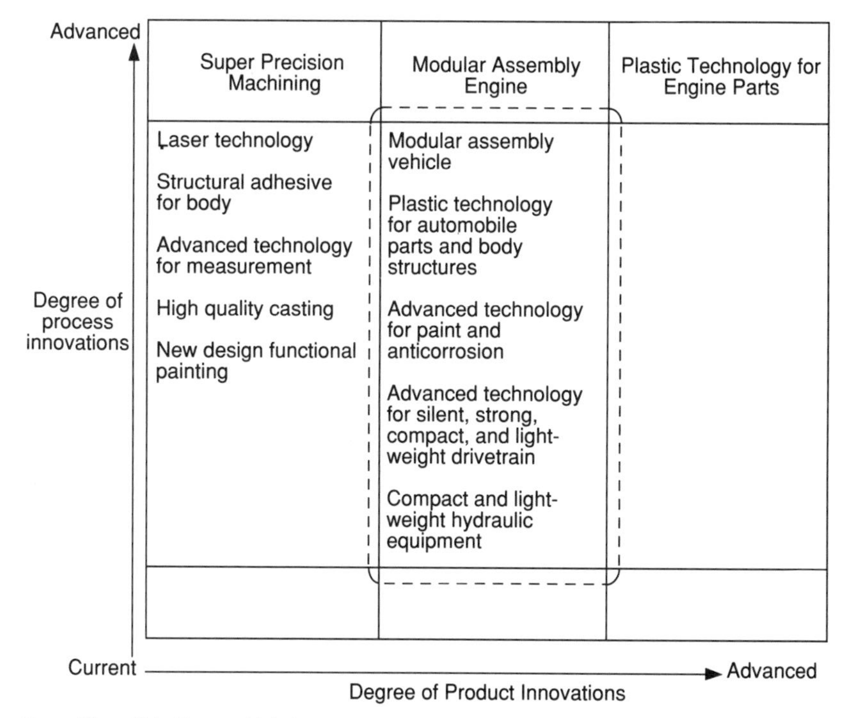

Source: Nissan Motor Company Limited

Figure 3-6. Nissan has set up a concurrent engineering center in Japan to exploit new technology and new processes in all its models.

— one-third of the way through this particular design project. The plant was built in less than 6 months, and the manufacturing equipment was developed over a 12-month period, including the time the plant was being constructed. Improvement of the manufacturing process continued for another 12 months, during which time pilot production was undertaken.

These controllers consist of sheet metal housings and electronic assemblies. Matsushita operates on a six-day order-to-delivery cycle, with the first two days being taken up in sales and administration. On the third day, the CAM (computer-aided manufacturing) data are verified and the schedule is determined. Production takes place on the fifth and sixth day, and the assembly is shipped immediately. This speedy response is another facet of the Japanese approach that is setting the standards in terms of one aspect of customer satisfaction.

The Japanese have a long history of use of CE elements such as QFD and Taguchi's methods. Many have built their whole approach to product development around QFD and cross-disciplinary teams. Honda evolved its SED system to meet a particular situation. As the speed of its product introductions and new ranges of products shows, Honda has gained enormously from the use of the system.

Nissan has attempted to pick up CE in a form similar to that adopted in the United States, and should soon begin to reap the benefits since its manufacturing system is well suited to shorter lead times. Many other Japanese companies combine their use of groups to tackle new product development with some elements of CE, particularly QFD, but always with a group approach.

Japanese Methods Exportable

Some managers doubt that Japanese management methods can be adopted in the United States. After all, the Japanese culture and work ethic are quite different from that in America; the Japanese are largely a homogeneous race, and they have lifetime employment and tremendous company loyalty. To them fairness means being fair with other insiders — whether of the same company or the same family — not with everyone. The differences are substantial.

Nevertheless, there are many common factors among manufacturers in the East and West. Many Japanese techniques for managing manufacture came from the United States — two examples being QC circles and W. E. Deming's philosophy of quality, which have been adopted widely by the Japanese. In any case, there is considerable evidence to show that as long as the culture of the employees is taken into account, Japanese methods can be adopted and used successfully elsewhere. TQC, JIT, QC circles, and indeed CE are flourishing outside Japan, in each case with the same aims but with slightly different methods. Nissan stresses that its approach is its own and that it will not work in that manner for every other manufacturer. Each manufacturer has to plough its own CE furrow.

Many companies in the United States and Europe are adopting CE so that they may overtake or at least stay competitive with Japanese companies. Most are convinced that CE is the only way to develop new products, while others consider it an indispensable tool for catching up to the Japanese. Nowhere are managers saying that CE is just another buzzword — something picked up, played with for a few years, and then discarded. All users seem convinced that it represents the future.

CHAPTER FOUR

North American Pioneers

Automobile industry's five years' experience in CE
CE to be obligatory for military equipment contractors
Digital and Xerox turn decline into leadership

The success of the Japanese in producing excellent products with short lead times made the U.S. automobile industry realize it had to make major changes in the way it operated to remain competitive. After investigating the problem the solution was born: concurrent engineering. The Big Three — Chrysler, Ford, and General Motors — started using CE in the mid-1980s. But CE was not used fully and took hold in only a few divisions at first. Indeed, the sheer size of the organizations presented major obstacles to the adoption of CE across the corporation.

At Ford, some executives were so shaken by the efficiency they saw in Japanese plants in the early 1980s that the company adopted an "after Japan" program of ways to catch up. Although it was Toyota and Nissan that were causing most concern, Ford's 25 percent shareholding in Mazda gave executives an inside view of the way the Japanese really operate. In addition, Ford became a customer for Mazda transaxles, and has since worked with Mazda on a number of joint projects, such as the current Escort and Probe.

Despite this early move into concurrent engineering, the automobile industry did not obtain good results in all projects. One reason was that task forces did not fit easily into the structure of the corporations, a situation exacerbated by the sheer size of the team needed just for design in a program to develop a completely new model.

There were also periods of disillusion; some found that once a vendor for capital plant was involved in the task force, it seemed to grow less cooperative. In turn, vendors complained they were not being treated as real partners and that they were expected to make substantial changes to the machines at their own cost just because they had a person in the CE team. They felt that despite CE, late changes were still being made.

Today a more realistic approach is being taken to CE, but the important point to note is that despite five or six years' experience, CE is spreading through the different divisions of the automakers and some of their vendors. CE is still seen as the main thrust to improve quality and cut lead times; the industry sees no alternative weapon. Rather than abandon CE, automakers are changing their management structures to suit CE, and are integrating it into other policies.

That CE is growing rapidly in the United States is supported by a recent survey that put the market for hardware and software for CE projects at $18 million in 1990, growing to $90 million by 1995. That sort of growth is clear evidence of the enthusiasm among American manufacturers for CE. In any case, these figures understate the case because they refer to computer equipment and the hardware of manufacturing itself.

In 1985, Chrysler established the Liberty Team, in which product design and manufacturing engineering work together on new ways of producing vehicles. This led to a comprehensive approach to CE. Ford actually came into CE through extensive use of DFMA on the Taurus program, and then moved on from there. At GM, CE was an integral part of the Buick Riatta, 1989 Cadillac Fleetwood and De Ville, and the Lumina MPV. It is now used for all the company's engine programs.

In the electronics industry, Xerox and Digital Equipment are both committed to CE, although as mentioned earlier, Xerox, which has spent eight years refining its system, does not call it CE. Both have made significant gains as a result of their approach.

The CALS Initiative

It is not only in the automobile and electronics industries that CE is coming into use. Some aerospace companies, such as Northrop, are also adopting CE. Although the manufacturers can see substantial benefits in its use, they have another incentive: the fact that the Department of Defense (DoD) is demanding its use. This is a most dramatic move, and one that will have far-reaching effects throughout U.S. industry.

The approach specifically demanded is the CALS (Computer-aided Acquisition and Logistic Support) program promoted by DoD and the Department of Commerce. The ultimate aim is for all new weapon systems to be developed in a CE environment and for all the technical data, including manuals, to be held in a distributed data base. All U.S. armed services, federal agencies, and contractors and sub-contractors to the Pentagon will need to adopt the CALS approach. In addition, many companies in associated industries, such as commercial aircraft manufacturers, and their vendors are being drawn into the CALS net. Thus it promises to help upgrade the efficiency of product design and development across the nation.

In combining the concept of an electronic data base with CE, the DoD has a simple target — it wishes to cut the cost of its weapons over their whole lives, and to reduce lead times. In fact, the DoD talks of *concurrent engineering* or CE, rather than

SE, while the Air Force refers to *integrated product development*, or IPD, but despite the different terminology, the language is the same.

The DoD is among those organizations that have studied the potential gains of CE and found that impressive benefits are obtainable. Although these are in line with those quoted earlier, they are especially significant coming from such an organization and because they refer to complex systems produced in low volumes. The DoD found that with CE,

- design changes can be cut by 50 percent
- development lead time can be cut by 40 to 60 percent
- scrap and rework can be cut by 75 percent

Since most weapons remain in service for 20 or 30 years, it is vital that the latest data always be available for maintenance and for any modifications required. Evidently, the sheer size of manuals that each unit of the army needs to carry to maintain its vehicles, weapons, and tanks is enormous. Moreover, the updating of manuals in the field is rarely thorough and often takes place a long time after it should. With an electronic data base, the up-to-date data will always be available, although the military personnel will need to carry portable computers to download the data.

CE a Cornerstone

CE has been made a cornerstone of the CALS project in an effort to eliminate the dreadful record of late delivery and cost overruns that are typical in military contracts. Most weapons are heavily modified late in the program, at considerable additional cost, and are therefore delivered late. Since we have enjoyed peace on most fronts for a long time, the lateness to market has not been too serious, but the high cost has been a major problem. Despite the success in the Gulf, the war there concentrated the minds of the military on the seriousness of late delivery of new weapon systems. It also showed that the U.S. military needs to be able to move huge amounts of equipment long distances in a short time, and needs to be able to maintain and replace them in difficult environments.

If the DoD is to provide the military with the latest weapons and systems — some of the electronics that played a vital part in the Gulf War were developed in the 1970s — costs and lead times must both be slashed. In addition, the DoD needs to be able to update weapons with improvements when they are needed for military reasons, not when it is practical to make the changes logistically.

Therefore, as in other CE projects, CALS will require a much fuller definition of each product at the concept stage, and the discipline of accepting that the definition is complete will fall just as heavily on politicians and top military personnel as it will on the manufacturer. Subcontractors will need to be involved in the multidisciplinary teams, particularly those supplying critical components such as landing gear and engines of fighter planes.

Policing Changes

Compared with industry, the DoD will have a much tougher job in policing changes. To start with, projects take far longer from concept to production — currently 7 to 15 years. With CE, these lead times should be reduced to 3 to 8 years by the end of the century, but even then a major review of the design is almost inevitable halfway through the development period.

In addition, developments in the world's power structure, and new intelligence regarding the weapons of potential enemies will put pressure on the project managers to change the specification in some way. Changes will need to be made at the concept stage or, in the case of very long projects, at prespecified review meetings. Military equipment lends itself to modular design, with the cut-off date for the design specification cascading down from the concept through the major units to the smaller assemblies. So there are natural points at which changes can be made to the basic structure and its modules and subsystems.

Pentagon staff members need to be able to judge future trends well enough in advance to allow for changes to be practical and for their introduction to be phased in as they would be in industry — when the design is updated. Once CE is in use, the late changes that push a project back two years and add $100 million to the budget will not be acceptable. Obviously, a new way of thinking is required to handle changes while keeping the military equipped with the most modern equipment available.

EDI and CE in Phases

There are two phases to CALS, and the first phase started in 1990. That involves the storage and transfer of engineering data by electronic media. Thus, when a weapons system is developed, the contractor will be obliged to submit engineering drawings to the DoD by EDI (electronic data interchange). The vendor will need to maintain its data base, and allow the DoD to access the data it requires. This approach cuts straight across the wall of confidentiality that contractors like to keep between themselves and the Pentagon, and is not being accepted easily.

The DoD wants the data base to be used for all information on the product, including manuals. It envisages the maintenance crew on a gun in the desert accessing the data base via satellite to obtain the few pages of instructions it requires to replace a broken spring, for example.

Although the adoption of CE principles in project management will be difficult enough, the need to take into account whole life costs rather than manufacturing costs will be difficult for vendors. The Pentagon is reviewing its requirements in the life of weapons, and the Gulf War has no doubt led to some priorities being changed already. The DoD's goals are identical to those of CE:

- Better quality
- Shorter lead time
- Products designed to last for their complete life
- Lower costs

Within the next decade, most companies that supply defense equipment or are subcontractors to that industry, along with the manufacturers of electronics and a host of other equipment, will find that they are being asked whether they have adopted CE either by the DoD or by their OEM (third-party developer) customers. To avoid becoming involved in CE under this sort of pressure, manufacturers should take the step now.

One corporation involved in aerospace and defense equipment that has adopted CE is Northrop Corporation. It has spent six years developing a system for use in its products, and gained significant benefits in recent programs. For example, internally generated changes have been cut by 75 percent, and changes in the validation program have been cut by 46 percent. Work-in-process has been reduced by 50 percent. Boeing has also adopted CE in the development of the 777 airplane, with co-location of the design and manufacturing teams, but it is too early to talk of the benefits.

Another company that has adopted CE partly as a result of its involvement in the defense industry is Smiths Industries Aerospace and Defense Systems (Grand Rapids, Michigan). It manufactures electronics systems for fighters and missiles, as well as flight recorders for civil airplanes. Responding to the impetus coming from its customers, Smiths Industries decided that it should adopt CE to improve its performance, and it sees CALS as enabling it to do so. Although a relatively new user of CE, it has already cut the lead time involved in carrying out major engineering changes by 25 to 50 percent, and it is now using CE on its bigger projects. This is typical of the approach in the defense industry.

From Laggards to Leaders

Those that do make the move to CE will find that there are plenty of role models to follow. Some of the greatest success with CE has been achieved in the electronics industries. Two outstanding examples are Xerox and Digital Equipment, both of which needed to make urgent changes if they were not to lose their position in their markets. Significantly, both corporations operate worldwide, with manufacture in several areas; thus they were, and are, closely in tune with world markets.

Xerox, which had milked the cash cow of xerography for well over a decade, found itself losing market share dramatically in the late 1970s. The market was being lost to a number of Japanese companies such as Canon and Ricoh, and the information coming back to the corporation suggested that poor quality was the main problem. Faced with this situation, Xerox decided improved quality was of paramount importance and that it therefore needed to adopt a new technique for developing products. Initially, only a few of the concepts of CE were taken on board, but as problems were encountered the system was modified; it is now as comprehensive an CE system as any. And it is still being modified, with new training modules being added — proof that once you move to CE you will stick with it.

Digital Equipment was also in a plight when it first took up concurrent engineering; it needed to do something drastic if it was to turn its entry into the workstation market into a success. When it decided to enter the market, it had assumed that a good product based on the latest technology would have a long life and be profitable. That proved to be completely wrong: the demands of the market were continually changing, so product life was short. It also found that more derivatives were required, and that the 30-month lead time resulted in a poor return on investment and a late break-even point.

To overcome these problems, it adopted CE in 1986-87, and as a result came back to take a substantial slice of the market — and make money. So, here are two examples where lagging companies were turned into leaders with the aid of CE. That both are large corporations is proof that the culture of an organization can be changed as long as the senior management has the will and imagination to do so. In both these cases, the spur to the use of CE was not just the recognition of serious problems in the marketplace but a determined effort to turn the tide.

To the Top of the Class

Xerox started on its *product delivery process* (PDP), which is CE under another name, in 1980 and has modified it many times since. Its situation was one that frequently occurs in corporations that grow rapidly with a tight grip on new technology. Xerography changed the way in which businesses operated, and so the market grew at a tremendous rate. In such organizations, growth tends to race ahead of the systems needed to support the enlarged corporation. Xerox was suffering from its success. Design was lacking the discipline of systems, a situation exacerbated by the fact that even now, the xerography process is not understood in complete detail. Also, there was not enough reliable information coming back from the customers — despite the fact that the copiers were normally leased.

Xerox had placed itself in a classic situation: the product was rushed into production too quickly in order to match competition, and as faults were encountered by the long-suffering customers, the changes came through. In fact, every single component on its main office copier product of the 1970s was redesigned after the machine had been introduced to the market. To update the machines, Xerox had a huge army of field engineers — not service engineers — who went out visiting customers to replace the poorly designed components with improved versions. Of course, because the copiers were mostly leased at that time, the cost was borne nominally by Xerox.

Senior management at Xerox realized there was no easy way to rectify this problem, improve quality, and regain their lost market share; the competition — by then almost entirely from Japan — was too good. They made good quality and customer satisfaction top priorities, and that led to the adoption of the product delivery process (PDP), which is now concurrent engineering through and through.

Xerox has used CE to pull itself up so that on its measure of customer satisfaction, the average for its products has increased from 50 percent to 90 percent; the corporation has a target of 100 percent for all products by 1993. Customer satisfaction in this context is the sum of answers to standard questions customers are asked. It does not mean that nothing fails, but it is nevertheless an important tool for finding out the customers' requirements.

Typical of the products developed by PDP are the 5028 copiers, one of the big-volume products. Because of problems the corporation had encountered with designs that could not be made easily, one aspect of the team approach in the development of that machine was to force design engineers onto the shop floor, somewhat along the lines of the Honda techniques mentioned in Chapter 3.

With the aid of PDP, time-to-market has been cut 30 to 40 percent, while some 75 percent of products that go through the concept stage reach the market, a far higher figure than was the case previously. This is of major importance, of course, since when a product is abandoned, considerable design effort is wasted.

Significantly, Xerox operates PDP on a worldwide basis, which includes Rank Xerox, its joint venture in the United Kingdom, and Fuji-Xerox, the Japanese joint venture between Rank Xerox and the Fuji group. The system is highly structured and built around a number of modules — over the years the system has been modified many times — but it is extremely close to CE in concept. The most recent modules cover aspects of product delivery such as the launch onto the market.

CE or Bust

Digital Equipment's problems resulted from changes in the computer business in the 1980s. Around 1980-81, customers were eager for any new technology, but toward the end of the decade they became reluctant to buy technology for its own sake. Buyers found that promised gains were not being achieved, that start-up problems were excessive, that the machines were too complicated, and that they often had to spend a lot of time and money on new software and updating their data bases. They started to doubt manufacturers' claims much more than in the past, and became reluctant to buy new machines.

At the same time, technological development continued apace so that if the development period was too long, it was very difficult to estimate exactly where the market would be when the product entered production. The product development cycle of 30 months was so long that it was crippling Digital Equipment. Typically, the life of a product was only 24 months before changes were needed. But the situation was worsened by the changes that would take place during development, so that by the time it had put a new workstation or minicomputer on the market, either extra derivatives were needed or the life was so short the product hardly broke even. Digital Equipment therefore adopted CE across the company's operations, and in three years had made the following improvements:

- Time-to-market reduced from 30 to 15 months
- Product costs cut by 50 percent
- Sales increased by 100 percent
- Break-even point reached earlier by about 6 months

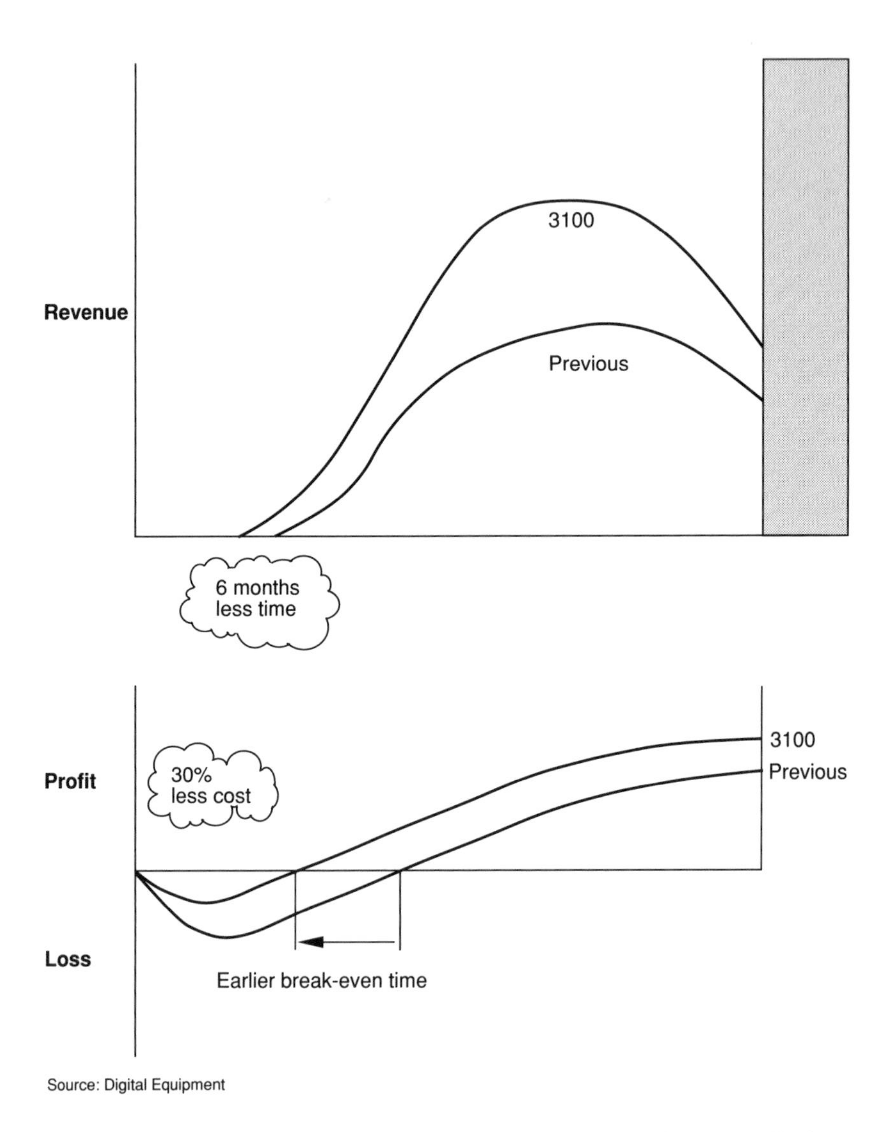

Source: Digital Equipment

Figure 4-1. Digital Equipment made substantial gains when it applied CE to its new range of Microvax 3100 machines.

- Profitability increased to 9 or 10 times what it was five years previously

All these gains came from CE, and now that the time-to-market is down to 12 months, the target is 9 months. At Digital Equipment, time-to-market does not mean from concept to Job One, but to the point at which planned production volumes are achieved. This approach was adopted because it was found that unless production was ramped up quickly, the break-even point was pushed far into the production run — with potentially disastrous results if the product became obsolete before planned. By contrast, in many companies, where product lives are three to five years, the production volumes in the first 3 months are pathetically low and then the buildup starts so that often the average production level in the first 12 months is below 50 percent of planned volume, which has an adverse effect on profitability.

In the development of the Microvax 3100 minicomputer, which is a complete range of products with greater variety and much more power than the preceding models, Digital reduced the cost by 30 percent and the time-to-market by six months. The product produced revenues earlier than the previous model and therefore increased market share, increasing revenues again. The result was an earlier break-even point, and greater profits.

Automobile Industry Leads

In the U.S. automobile industry, GM began using CE in 1984, and Chrysler and Ford started around the same time. One of the first GM projects was GM33 for the development of the Buick Riatta, a coupe. Because the Riatta was to be built in low volume, GM turned to outside consultants for much of the body design

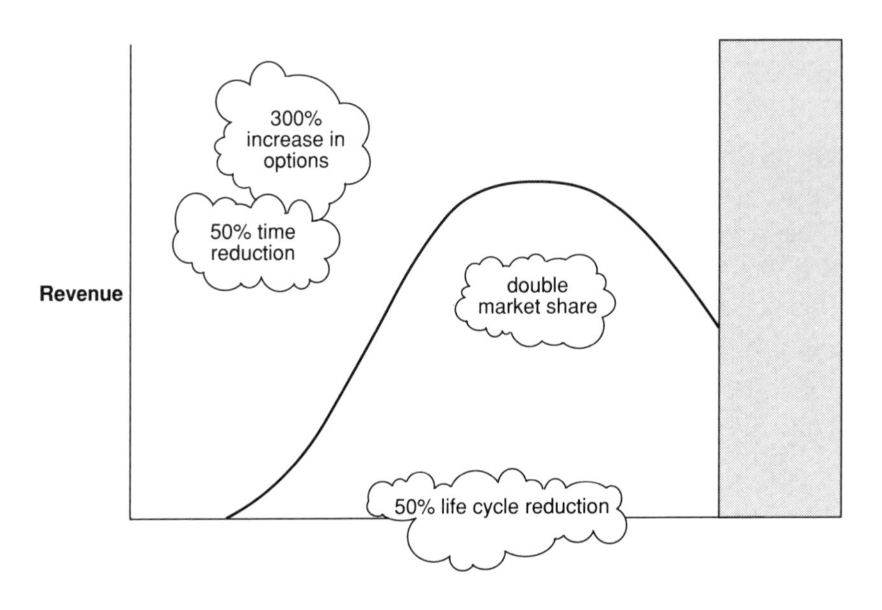

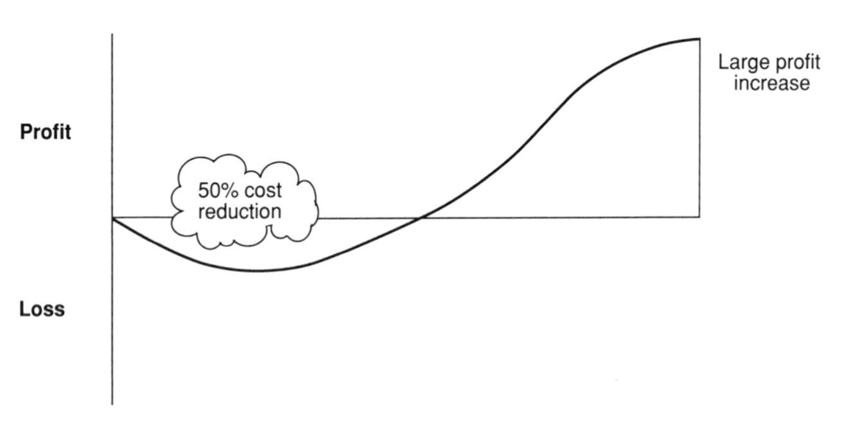

Source: Digital Equipment

Figure 4-2. With a shorter lead time and ramp-up to full production, CE brought increased profits from Digital Equipment's minicomputers.

and prototype manufacture. Hawtal Whiting, Inc. (Troy, Michigan) was responsible for the basic body design and prototype manufacture, Aston Martin Tickford (United Kingdom) performed the trim and final assembly, while Lamb Technicon was responsible for the body welding fixtures. Ogihara Company, Ltd., a Japanese specialist, was responsible for the body dies and tooling, so the project was CE on an international scale.

All involved were enthusiastic about the approach, with Lamb Technicon's personnel able to talk directly to those from Ogihara and know exactly how far the tools had proceeded at any time. If they thought a change would improve welding assembly, they could discuss it immediately. Such a situation would have been impossible with over-the-fence engineering.

For example, there would not have been one engineer totally responsible for the dies; rather, one manager would have had to consult the production supervisor and then go to the die designer to see what changes could be made. Everyone involved would have had to stop the work they were doing and start thinking about the GM33 project. Such switching from project to project means that people need extra time to orient themselves to the job, especially where intricate designs are involved. By contrast, with CE, every request for a change to the die was handled instantly, with all the people concerned consulted at once. Likewise, there was close cooperation between the body designers and tool designers. As a result of the use of CE, the project was completed in three years.

Chrysler's Team and Tools

Since the establishment of the Liberty Team, Chrysler has expanded its CE methodology to a complete company-wide system.* It was able to benefit from its comparatively small size

---

* J. P. Hinckley, Jr., *Early Designs for Manufacturing Quality,* SME Technical Paper MM88-151, 1988.

to move nimbly to the new methods. It has adopted the multidisciplinary team concept, QFD (Chapter 8), DFMA and FMEA (Chapter 9), and added a few concepts of its own. However, rather than adopt a neutral stance over the design/manufacture conflict, Chrysler defines the manufacturing process as the first step in the development of new models; its products are process-driven rather than design-driven.

Optimal manufacturing processes are planned and used to generate guidelines for the product design. Although this is a sound approach when the investment in manufacturing plant is a major constraint, it is not suitable for all products, since it will tend to eliminate those radical solutions that combine an exciting design with a new process.

Chrysler also operates a Product Confidence Index for each model program. This is approximated at the start and updated regularly. Thus, at each stage in the program, management can see where the potential problems will come. Members of the task force concentrate on raising the index to acceptable levels, and know that if the index is not high enough, the product cannot be signed off.

Another significant tool is the simulation of assembly, taking into account the expected variations in dimensions. This technique highlights and ranks potential problems. In addition, of course, it shows the effects of changing any tolerances.

That CE can be used to make gains outside the narrow confines of product design is shown by Chrysler's adoption of modular concepts to reduce parts handling in the plant. Currently, each assembly plant handles 3,000 to 3,500 discrete components and subassemblies. Chrysler has adopted a program to cut the number to 1,500. Some gains will come from the reduction in parts count that follows exhaustive use of DFMA, but a larger proportion will come from the use of modules to be subassembled by vendors or off-line. Among the 28 modules to be treated in this way are the instrument panel, door assemblies — already assembled off-line in the majority of Japanese plants — and some major chassis units.

Cadillac Cuts Lead Time

When Cadillac was reorganized as a total car company in 1987, it expanded CE throughout all operations, although it had already been using the approach in limited ways before that. In fact, the completely redesigned 1988 Eldorado was taken through its complete program from start to arrival in the showrooms in 55 months, compared with the 75 to 80 months that had been normal at Cadillac. In addition, minor redesigns — halfway between complete redesigns — now take 12 months instead of three years. Cadillac has developed a comprehensive system to manage CE.* There is a steering committee that reports to the company's executive staff and attempts to optimize the use of CE. Each vehicle program is handled by a vehicle team, a truly multidisciplinary group that includes

- vehicle chief engineer
- design staff
- manufacturing and manufacturing engineering
- materials management
- sales and service
- finance
- market research and analysis
- strategy manager
- human resources
- customer satisfaction
- program coordination
- public relations

* R. H. Walklet, "Simultaneous Engineering: A Cadillac Perspective," in vol. 1 of *Proceedings of the 21st International Symposium on Automotive Technology & Automation (ISATA)*, Wiesbaden, November 1989, pp. 69-79.

The team is responsible for all aspects of the vehicle, including matching the business plan. As the approach was developed, the company created six vehicle system management teams (VSMTs). These are

- chassis/power train application
- exterior component/body mechanical
- seats and interior trim
- electrical/electronic equipment
- body-in-white
- instrument panel and heating/air-conditioning systems

Each of these groups is responsible for the systems it engineers and plans for all vehicles in the Cadillac range. Each VSMT is a multidisciplinary team with the same makeup as the vehicle teams.

In addition, there are product development and improvement teams (PDITs) responsible for the actual design of components in the six systems. They take cradle-to-grave responsibility for their components, and each PDIT is responsible for its components on all Cadillacs in terms of quality, cost, timing, reliability, and profitability. Essentially, each PDIT is a minibusiness. Cadillac has some 55 PDITs, with an average of eight members in each.

To gain the voice of the customer on a continuing basis, Cadillac has established a number of listening-post dealerships. Each dealership listens for complaints or opinions on a particular product group, and feeds information directly to the PDIT or VSMT. UAW members are also involved in the continuous improvement process, and part of their job is to visit dealerships and report back on their findings.

Cadillac is also keen for the team members to adopt a hands-on approach, so when there is a problem on the shop floor, members are encouraged to temporarily replace operators on the line and gain firsthand information on the difficulties being encountered. This is an excellent approach.

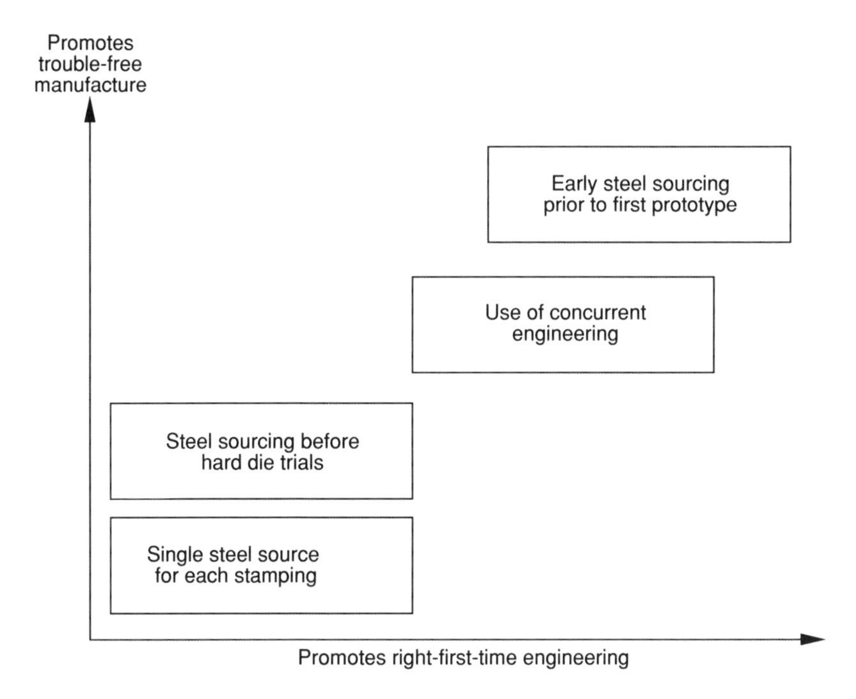

Figure 4-3. GM's Flint Automotive Division found that the combination of CE and the early involvement of the steel suppliers in the task force eliminated potential snags from manufacture.

Early Selection of Steel Vendor

Because of the complexity of form of auto body panels and the thinness of the material, the quality of the sheet steel has a significant effect on the quality of the steel stampings. The perceived quality of the paint finish depends on the steel quality, so the steel company can have a direct influence on the customer's view of a car. With this in mind, the Flint Automotive Division of GM involved the steel suppliers in the task force for the 1989 Cadillac Fleetwood and De Ville, one of the first programs for CE adopted comprehensively in GM.

GM emphasizes that the involvement of the steel companies represents GM's commitment to them but does not imply that they that have a contract. Several sources are still used, but the aim is for one source to be used for the complete production run for a particular stamping; the four chosen initially for the 1989 Cadillac program were National Steel, LTV Steel, Bethlehem Steel, and Inland Steel.

With this policy on sourcing, GM gains the benefits of single-sourcing in terms of consistent quality and accountability without committing all the production of one car to one steel maker. In addition, it is able to optimize the steel for the form of the panel. Panels that are deep-drawn require steel of different characteristics from large, almost flat panels, while in some areas high-strength steel is used.

Once the commitment was made, the representative of the steel company was involved in all relevant task force meetings and was expected to give any assistance needed to lead to a production-friendly design. Although it is always useful to involve the steel suppliers early, GM found that this was impractical outside the CE environment, because verification of the styling concept came too late and too many changes were being made after the specification of the sheet steel had been finalized. Without CE, it is practical to finalize sourcing coincident with hard die trials, but that is as far as the company can go.

On the 1989 Cadillac program, six steel companies were involved in CE meetings two and a half years prior to the start of production. Six months later, final decisions on sourcing were made, and since then this concept has been used on all models at Flint. This lead time is necessary to ensure that prototypes can be built from steel that is typical of the production material. The next step is to reduce the number of suppliers of steel sheet, with a target of only two for each car model.

In the power train division, significant gains were made in the design of an automatic transmission for a GM truck with CE. Since the production engineers and foundry experts were involved from the start, it was decided to replace the conventional one-piece housing, which is extraordinarily complex, with three housings. The result was that the impregnation, normally needed to combat porosity, was eliminated along with several machining operations. With these changes, GM

- reduced material cost by 50 percent
- reduced scrap by 60 percent
- reduced the number of tools required in machining by 30 percent

This is typical of the gains that can be made, and shows how North Americans are regaining their ability to cut costs and introduce models as quickly as the Japanese.

GM Truck & Bus (Pontiac, Michigan) has also taken CE on board, with the GMS Sierra and Chevy CK pickup trucks among the earlier projects handled this way. An example of how CE overcomes production difficulties is shown by a potential conflict over a tight tolerance on panel thickness where a tolerance stack-up problem had been encountered. Product design specified a tolerance of +/-0.0005 mm to overcome this stack-up problem in the cab. The manufacturing engineers considered this to be unacceptable because of the tolerance on sheet steel coming into the plant. Discussion by the team resulted in a practical tolerance and the elimination of the tolerance stack-up problem.

CE for the Lumina MPV

CE was also adopted by GM in the design and development of the Lumina multipurpose vehicle (MPV), of which Pininfarina of Turin, Italy, was responsible for the body styling. This was an unusual design with plastic panels hung on a steel frame — as on Saturn — and with planned output at the low level of 160,000 a year. Therefore the design and development program might have been expected to be longer than normal; in fact, the project took three years, against the five years that had been taken previously for such projects.

When the project was started in 1985, GM had no real procedures for CE, and these were developed as the project progressed. There was some nervousness within GM about the project because there were only 65 people in the task force, whereas it was normal for about 400 engineers to be involved in a project of this type.

Comau Productivity Systems was the supplier of the Robogate welding line for the Lumina and acted as liaison with Pininfarina for much of the work. It also produced the inspection jig used to check the plastic panels. The project's success owed largely to the close liaison between all involved, and the fact that the bulk of changes were made when the product was still at the design stage. Although the task force worked together much of the time, it met with the finance department people monthly to review the budget.

Ford Starts with DFMA

Ford's involvement in full-scale CE came through the Taurus project, the first model in which DFMA (see Chapter 8) was used throughout. It became apparent that a multidisciplinary team was needed to extract the full benefits of DFMA. Although the Taurus was not a full CE project, the use of multidisciplinary teams did overcome what might have been major problems otherwise. Board approval for the Taurus and Sable was given in the spring of 1980, and at that time the plan was for a fairly compact car. However, by April 1981 it became apparent that the market requirement was changing, partly as a result of falling oil prices.

It was realized that the car was likely to be more successful if it was made bigger all round — wider and longer, with a longer wheelbase. Normally, such major changes would have resulted in a serious delay to the project, but because of the team approach and DFMA, the delay was minimized. Ford has also adopted CE for some of its engine programs, such as the 4.6 liter V-8, discussed in Chapter 11.

Other U.S. Examples

Among other U.S. companies using CE are the power tool maker Black & Decker (Towson, Maryland) and Remington Arms Company (Illion, New York) which adopted a multidisciplinary team approach, involving engineers from its parent Du Pont, when it introduced a flexible manufacturing system (FMS) into its plant.

Delco Remy, which supplies about 30,000 sets of electrical equipment daily to the automobile industry, has also been using CE for several years.* Its first project was in the development of batteries for automobiles, and the team includes representatives of the hourly-paid operatives on the line. Delco Remy found that minimal bureaucracy was essential if the task force was to be successful, and that the practice of standing together on the line or in the development shop observing a problem gives the product designer and manufacturing engineer a greater understanding of each other's difficulties.

Delco Remy has used CE on a number of projects including starter motors, alternators, and control valves. In one case, the cast housing could not be produced in high volumes as designed. The team, which included representatives from the

* D. D. Barron, *Simultaneous Engineering at Delco Remy,* SME Technical Paper MM88-154, 1988.

vendor of the casting and from purchasing, was able to implement a redesign to improve producibility without adversely affecting the performance.

In another case, the switch from the use of a deep-drawn housing to an extrusion allowed the use of a smaller and more rugged solenoid. The reason for this improvement was that the extrusion was produced with a very close tolerance, whereas the deep-drawn housing not only had a slight taper but had wider tolerances. This is one case where the improvement could come only with the team approach; use of DFMA alone would not have created the knock-on benefits.

Significantly, Delco Remy found that the biggest area of improvement resulting from CE was product quality, "because all the engineers involved were aware [of] and understood the requirements of the customer." As has been emphasized already, CE is all about improving quality and moving the responsibility for quality from police on the line back to design.

CHAPTER FIVE

CE in the European Automobile Industry

German automobile industry takes the lead
Rover Group cuts lead time in half

In Europe, the Volkswagen Group, Daimler-Benz, Ford, GM's subsidiary Adam Opel, Lucas Industries, Perkins Engines, Robert Bosch, and the Rover Group are among those using concurrent engineering. German corporations led, with Volkswagen having taken CE on board in the most comprehensive manner. BMW and Opel are moving in the same direction, while Daimler-Benz has adopted a more limited approach.

In 1987, Volkswagen's management held a meeting at which it was agreed that all the systems in use within the company had to be rethought. CE emerged as the new approach. After some planning and reorganization the company settled on the use of task forces for new projects. Each team consists of at least one member from

- research and development
- manufacturing engineering
- purchasing
- finance
- sales

Volkswagen (VW) considers task forces essential, but believes that a manager of the task force would be more trouble than he or she was worth. Thus, all members work on an equal footing. There is also an CE steering committee that meets weekly to discuss problems and developments in the methods.

Interestingly, VW considered using a high-speed data highway to connect team members who would still remain in their own departments. A few experiments showed that this approach was unsatisfactory. The result was not CE, but sequential engineering. People would not pass on the data until they had finished their work, whereas one of the features of CE is that participants learn to work with incomplete data. Also, VW found that component "make or buy" decisions had to be made very early in the project. This required that staff from the purchasing department be involved from the concept stage onward.

The head of the department responsible for CE at Volkswagen research and development points out that "every day we have to think about changing the processes, about doing things differently." He adds that otherwise, people will soon slip back into their old ways.

Disbanded Quality Control Department

VW also took the major step of disbanding its quality control department and deploying the staff into departments where their experience could improve quality at the source. It decided that the department's existence allowed other departments to pass the buck on quality. Now, everyone is responsible for quality, with production taking control of statistical process control (SPC).

Because VW is located far from an industrial area, it makes more prototypes and prototype components than most companies. When CE was introduced, the company took advantage of its large prototype shop to involve production planning engi-

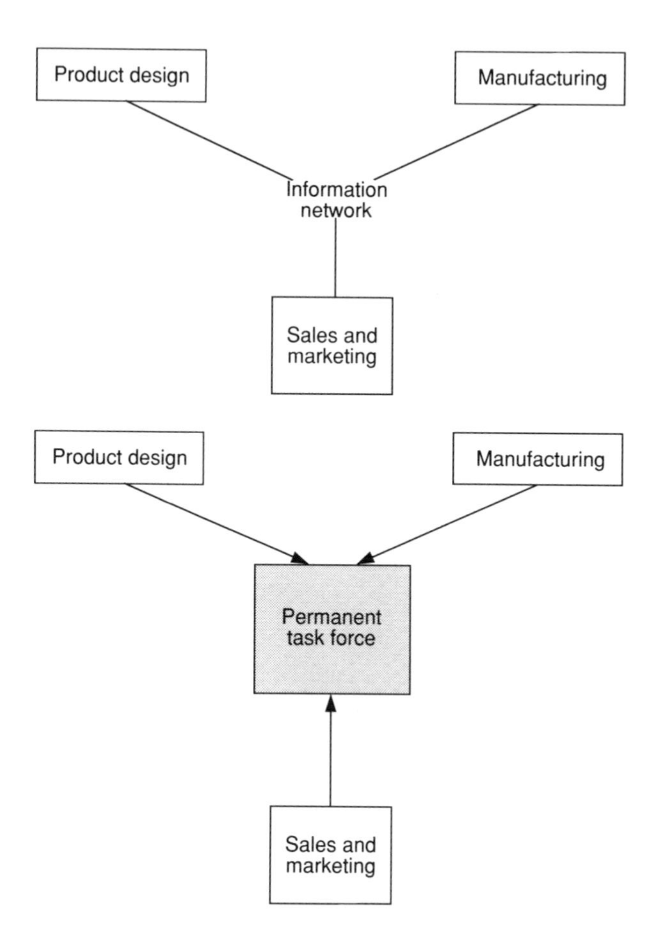

Figure 5-1. CE is not achieved if people from different departments communicate over a computer network; they must work together in a task force.

neering and personnel from the pilot production shop, which normally irons out assembly problems, in the prototype building program. The pilot build staff not only speed up prototype assembly, but they can also point to any impractical features well before pilot build takes place.

Before CE, VW also had some logistic problems in persuading product engineers and manufacturing to liaise closely; the manufacturing engineering department is 15 miles from R&D. Now, a new technical center is being built adjacent to research and development to house project teams and manufacturing engineering.

CAD for Simultaneous Product Design

VW puts much faith in the use of CAD in its approach, relying heavily on programs developed with Control Data (see Chapter 11) to move the concept forward simultaneously in product engineering. For example, stylists can be working on the body design while engineers iterate simulations of aerodynamics, torsional strength and other structural analyses, crash performance, and vibration analysis — almost all at the same time. The VW Futura design concept car was designed in this way, with the complete car designed using CAD. The first production vehicle to be designed totally by CAD was the new Golf Cabrio, while an engine was chosen for the first full CE project.

VW now claims to have reduced time-to-market for its new models from about 48 to 36 months. Its target is now 32 months. At the same time, it is confident that with continuous nurturing, quality will improve, while manufacturing costs — still a major problem — will be reduced. At the same time, Volkswagen executives state that the main problem in introducing CE is in persuading people to accept change — and that they have to work at this every day.

In many respects, the approach at BMW is similar to that of VW, except that BMW put the building of a new multidisciplinary research and development center first. Unlike the conventional R&D department, BMW's Forschungs und Ingenieurzentrum (FIZ) is the home for members from the following departments:

- Research
- Development
- Technical planning
- Production technology
- Quality assurance
- Value analysis
- Cost control
- Purchasing
- Logistics
- Patents
- Personnel

Teams of people from these departments work together on each job, and the prototype shop is next to pilot production. BMW has also spent heavily on computers to make possible the simulation of designs and functional properties with the installation of two mainframes and a Cray X-MP/28 supercomputer. It is using Catia and Exapt software for CAD/CAM and has 450 workstations installed.

Design and Production Engineers Together

Adam Opel, a member of GM Europe, is another German automaker that has taken up CE, but it has put the emphasis on making product design and manufacturing engineering work together. To this end, it set up a new department — Advance Product Study — consisting of product designers and manufacturing engineers. In addition, product engineering and manufacturing engineering now answer to the same director. As a result, the redesign necessary to meet the requirements of production — which took place three or four times with over-the-fence engineering — has been eliminated. All products that leave the Advance Product Study Department, which comes onto the scene halfway through the preconcept stage, are now guaranteed to be suitable for volume production.

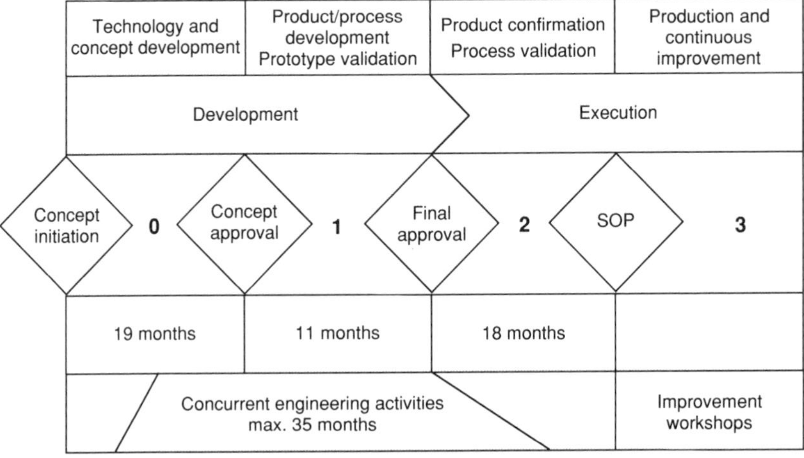

Source: Adam Opel AG

Figure 5-2. Opel's concurrent engineering activities are concentrated in the stage prior to final approval.

Nevertheless, a multifunctional team, with representatives from marketing, product planning, styling, product engineering, production engineering, purchasing, quality, service, and finance, is used for all new model programs. This team operates from preconcept stage into production, coordinated by program directors from the technical development center.

Within the overall program, there are teams for specific products — such as engine and body — and according to the relative importance of design and manufacture, either a product engineer or manufacturing engineer will lead the team. GM Europe has also trained more than 100 engineers and managers in quality function deployment, and the program continues, as the corporation gears itself for general use of the tool.

One of the first products to be developed by a team of product and manufacturing engineers in the Advance Product Study Department was a new rear suspension. Lead time was reduced from 14 to 3 months. Other projects have involved steering components and body design.

Robert Bosch is one of the European component vendors involved in CE. It cites the design of headlight reflectors as an example where CE, in this case heavily dependent on CAD, resulted in the lead time for a new product being cut from 24 to 12 months.*

Back from the Precipice

Another company that needed to claw itself back from the precipice leading to extinction was the Rover Group in Britain. It took a license for a small Honda car, then worked on some joint ventures with Honda; inevitably, CE was the result. Traditionally engineered projects at Rover were taking six to eight years from management approval to production — and often when production started, the output level was no more than pilot production in world-class terms for 6 to 12 months. Now, Rover engineering management is confident that it can take a project from clay to production in 48 months.

It has its sights set on three-year programs, and indeed, its Land Rover division took the Discovery four-wheel drive vehicle through its complete program in 27 months. However, since this vehicle has a separate chassis and body, and most mechanical units were derived from other models, it would not be expected to take as long as a normal passenger car. Nevertheless, the project, run by a team of 25 people all working in the same room, was highly successful, and sales of the Discovery have broken all records for Land Rover.

---

* G. H. Schoeffler, "A Supplier's Approach Towards Reduction of Product Development Time in Automotive Lighting by Simultaneous Engineering," in vol. 1 of *Proceedings of ISATA*, pp. 157-163.

In its passenger car division, Rover Group has yet to achieve the rapid ramp-up in production that the Japanese maintain. Indeed, it ascribed its loss of production in 1989 to the introduction of new models, something that the Japanese just do not let happen.

Rover started using quality function deployment (QFD) in the middle of 1989 for all projects. Honda also insisted that "the humdrum was as important as the exciting" and that no project could proceed beyond the concept stage before a thorough design for manufacture and assembly study had been carried out. Rover has now adopted these disciplines, and has taken up the Honda system of having guest engineers from vendors of significant components working in its design offices alongside its own engineers. Now, with three years of various aspects of CE under its belt, Rover is intent on achieving "world-class reliability and refinement" with the aid of CE to make its engineering more thorough, and to make it more responsive to customers' requirements and complaints.

Clearly, manufacturers in Japan, the United States, and some European countries in a variety of industries are already reaping the benefits of CE. The techniques are well proven but do involve a change in style and structure of management, and in relationships between vendors and manufacturers. But the benefits are so great that no manufacturer can afford not to use CE. As a senior engineer in a firm of consulting engineers that works for the automobile industry observed: "All our projects are concurrent engineering projects because that is the only way to carry them out efficiently on time."

PART TWO

Tools and Techniques

CHAPTER SIX

Management in Control of Product Development

CE cuts lead times
Reduces the number of late changes drastically
Brings the break-even point forward
Ensures quality from the outset

For concurrent engineering to be a success, senior management must be dedicated totally to making it work. For example, more investment is required early in the project than with conventional engineering, while a hands-off approach is needed — the task force is in complete charge of the project.

At all costs, senior managers must avoid trying to control the task forces too closely; the members of the team need to be able to make their own decisions based on the ground rules laid down by senior management. Needless to say, the board of directors will remain in control of policy and capital expenditure, and department heads will still control their own departments and their budgets.

In return for this hands-off approach, senior managers have everything to gain and nothing to lose except their previous frustration over the poor management of new product development and excessive number of late changes. They can expect to see products developed that are as close to market requirements as practical, have high quality, and enjoy costs that are under much closer control than previously. They can look forward to

- products that match precisely customers' needs
- shorter time-to-market
- earlier break-even point
- fewer changes late in the program, reducing cost of development
- simpler and cheaper manufacture
- assured quality from Job One
- low service costs throughout the life of the product
- less risk of failure than normal

These gains may seem pie-in-the-sky to a vice president who has spent his or her life in manufacturing; after all, they conflict with one another, and normally some are achieved at the expense of others. With CE they can be achieved and, in many cases, one advantage is made possible by an improvement in another. Integral to the concept of CE is the consideration of whole-life costs, which may result in increased manufacturing costs. The accent is now on value rather than initial cost, a concept that some senior managers may not accept easily.

However, there are also a number of areas where costs are reduced directly. For example, more testing is carried out on subsystems mounted on rigs than on prototypes in laboratories or on the road. In one case, development costs were reduced by more than 20 percent. Then, with conventional engineering, designers, manufacturing engineers, and other professionals can spend 20 to 25 percent of their time waiting for others to complete their work. With CE, the task force can soon avoid this problem, bringing savings of 10 to 15 percent.

Biases of Marketing, Designers

Conventionally, the specification for a product is defined by market research, with considerable input from product planning. Often, market research people slant questions so that they will receive the answers they want, or they interpret the figures

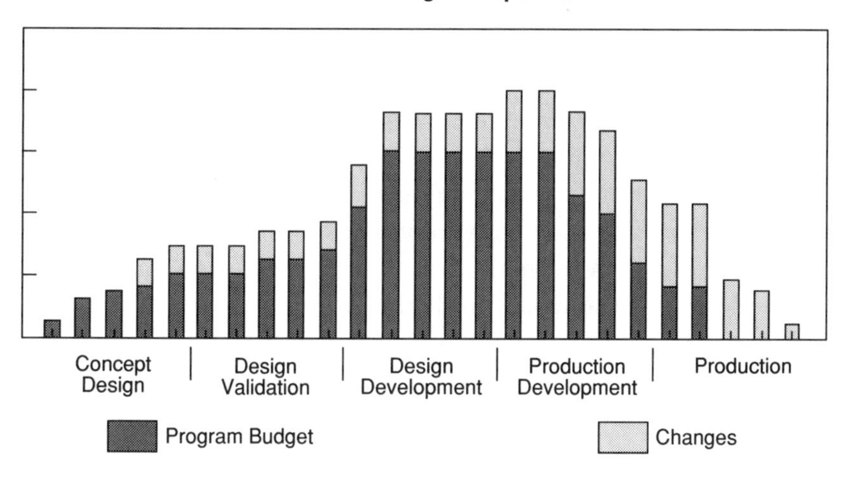

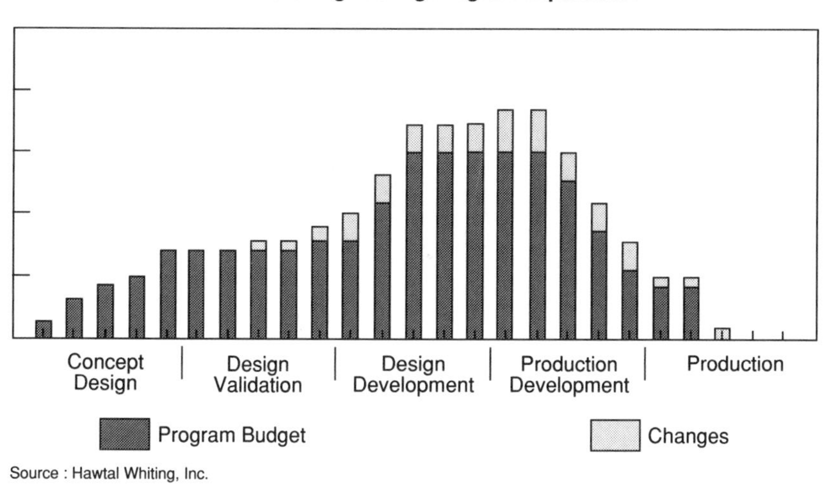

Source : Hawtal Whiting, Inc.

Figure 6-1. Senior executives will find that with CE, costs in each month are below those for a conventional program, as are the unpredicted costs that result from changes.

to meet their own expectations. As long as the marketing or sales department is responsible for gathering the data, the corporation is not hearing the true voice of the customer — it is hearing a biased voice. Very often, that bias is so strong that it represents the views of only a small proportion of the customers. For the true voice of the customer to be heard, the ears of the task force must be listening.

After this false start with conventional engineering, the product is specified from the outset in engineering terms, and engineers tend to design the product *they* think people want rather than what people actually want. It is only when the design has been virtually frozen that customer clinics are undertaken to discover what a few members of the public think of the product. Too often, these clinics are used to give management confidence in the product, with criticisms not being given the weight they deserve, largely because it is too late to make any major changes.

Adding Customers' Own Words

With CE, the product is defined by customers in their own language; that description is then turned into an engineering specification. This is an extremely important procedure, because the act of converting the description into engineering terms eliminates confusion or misunderstandings. In addition, it is necessary for all involved in CE projects to adopt a new approach to quality. To do this, personnel at all levels need to put themselves in the shoes of the customer; they need to act as the critical customer when in a shopping mall or restaurant and then translate those thoughts into their own product when they are back at the office. The need for this new approach is covered further in Chapter 14.

An exercise of this type reveals that the customers' thoughts on quality are very different from those of the manufacturer,

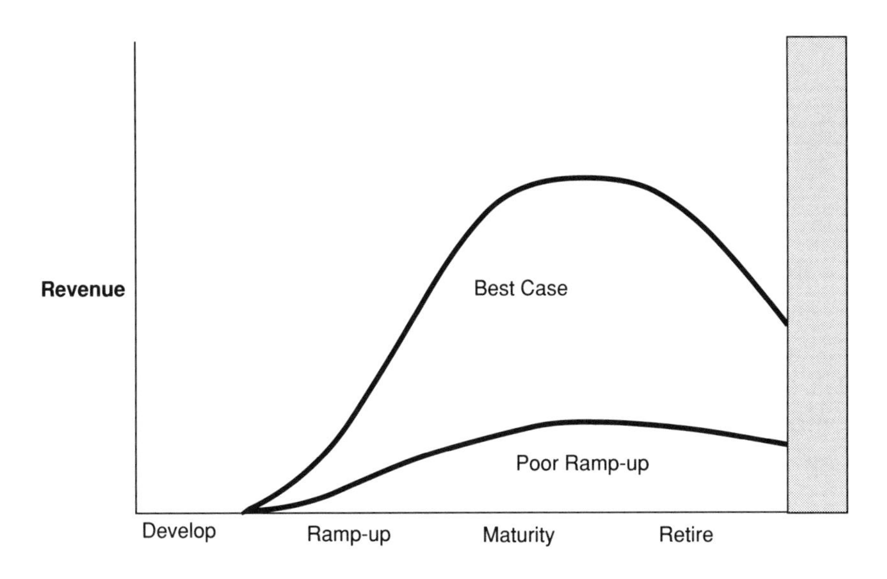

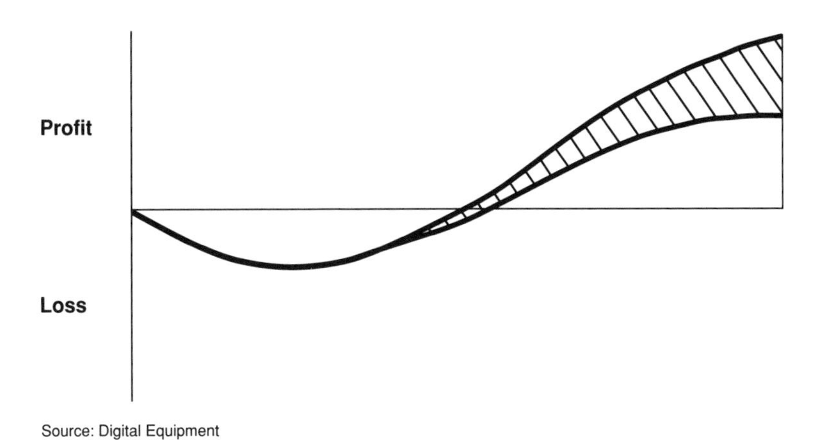

Source: Digital Equipment

Figure 6-2. The fast ramp-up in production achieved by Digital Equipment's CE program improved profitability dramatically.

who tends to think in terms of overall performance balance. An interesting example is in the finish of the paint on a car body. U.S. manufacturers have spent vast sums of money on refining the paint process to eliminate a characteristic called "orange peel," in which the paint finish is rippled. In the 1930s, when labor was cheap, operators would polish the ripples out. Now, the painting techniques have been refined to leave the minimum of orange peel.

In any event, when executives from the automobile manufacturers visited the Japanese automobile plants in the late 1970s — at that time the Japanese were beginning to build up substantial volume and take a sizable market share in the United States — the Americans were amazed to see a lot of orange peel. They presumed that the Japanese were unable to obtain a smooth finish in the short cycle times they had adopted and were unwilling to incur the cost of manual sanding of the undercoat, which also helps reduce orange peel.

Some years later it was discovered that this was not the case at all; the Japanese had carried out surveys on paint finish and had found that most customers at that time were happy with the orange peel finish because they thought it showed that the paint was thick and therefore durable. This survey saved the Japanese a lot of money and represents a golden rule of one of the techniques of CE: Do not give customers features they definitely do not want.

Since product requirements and fashions change so quickly these days, it is essential to translate a specification into a product as quickly as possible — but that does not mean rushing through to prototypes immediately. That approach, which has so often been the death knell of good concepts, inevitably leads to many changes being made to the design at the prototype and preproduction stages. It also leads to chaotic timing, with no one really sure when the product will be good enough for volume production. Indeed, it is not unusual for changes to be rushed through just as production starts.

By contrast, with CE, a considerable amount of time is spent in specifying the product and, because the specification is complete and well researched and planned, subsequent changes are reduced.

More Opportunities

With CE, the presence of people with different backgrounds in the task force makes specification of the product much more thorough. In the early discussions, the marketing representative may realize there is a niche market for a derivative of the product, or that with minor changes it could be sold to two or three other OEM customers. Normally, he or she would not have the opportunity to give any input until after the design and tooling had been frozen. Thus, the production of that derivative would be uneconomical, or the changes would push back the program, while the doubling of volume would create too many problems to be pursued.

At the same time, at the concept stage the manufacturing engineers in the task force are able to spot flaws in the design and can often suggest a way of producing certain components in-house. They will understand from the outset what degree of flexibility is required, and may be able to recommend a new process that will cut costs drastically. In both cases, the project gains immeasurably from their early involvement.

But can't these gains be made with better liaison between departments, and with continuation of the existing hierarchy? The evidence on this point is clear: Ignore the task force approach, and the gains will soon diminish.

The task force acts as a catalyst to CE, broadening the gains of the disciplines themselves, principally because the members are working as a team. Each is putting his or her professional experience into the project, as well as broader knowledge of the product and manufacturing. Thus, the "frustrated designer" who so angers product designers in conventional projects can give useful input on design without upsetting anyone. Also, of course, each member of the team is a potential customer.

Some desirable changes may affect several different departments. For example, one member might want a change to simplify manufacture, and the product designer may agree. The change may affect some other aspect of manufacture, such as die costs, tool design, or another component. The member of the task force with the relevant knowledge can voice his or her concern immediately. Then, a thorough investigation can be made, and the team will decide which approach is best. Without the task force approach, the first change would have gone through before the second aspect of the problem was discovered — when the tool vendor was asked to requote for the different die and submitted a highly inflated price.

Over-the-Fence Woes

The tales of projects that have gone wrong with over-the-fence engineering are legion. There was the case in which a large company, striving to reach new standards of flexibility and short lead times, adopted a flexible manufacturing system. The production engineers worked to the dimensions of the largest component.

Before the product was finalized it became necessary to lengthen the biggest model, with the result that it was too large to go on the machining center or on the special unmanned vehicles designed to transport the workpieces between machines. The product designers assumed that flexible manufacture meant that anything could be made in the plant, and did not consider size to be a limit. The result was that an extra machining center and extra unmanned vehicle were required for the largest com-

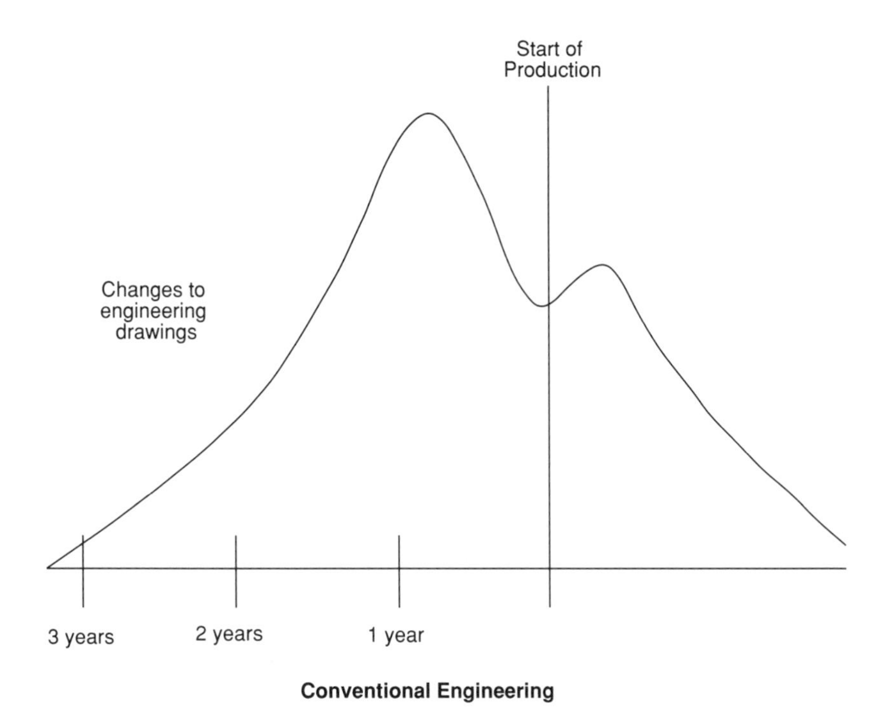

Figure 6-3. With over-the-fence engineering, too many changes come too late and at too great a cost.

ponents; these were not only the most expensive pieces of equipment in the shop but were underutilized. This situation in turn not only increased the cost of the plant but reduced operating efficiency as well.

Problems of this type will evaporate with CE, because production engineers have a much improved understanding of the product than previously, and the product designers now have a better idea of the limitations of the manufacturing plant. Any request to allow for a larger workpiece will come through much earlier than in the past.

Simulations in Parallel

With CE, the team investigates every facet of the product at the concept stage. CAD equipment allows many simulations on variations of the product to be run to determine factors such as structural strength, aerodynamics, and tolerance stack-up. At the same time — literally — manufacturing engineers can be running simulations of alternative manufacturing strategies, while key vendors can be putting components through similar programs. These simultaneous operations not only save time but allow changes to be made far earlier in the program than normally.

By contrast, with over-the-fence engineering, cost overruns for the machine tools and other manufacturing equipment are daily wrangles between purchasing and the vendors. The purchasing manager cannot understand why the vendor fails to stick to the price agreed, while the vendor is being forced to change the machine to suit product design's revised specification. The result is an immense amount of additional management effort at both companies — and it can all be avoided with CE.

A Little Extra Cost Now Saves More Later

The result of using concurrent engineering is that when the concept model is turned into a prototype, management has far greater confidence than ever before in all aspects of the product. Any components bought outside will have been produced by the potential vendor, which will also ensure that the vendor knows what to produce before production starts. Tools will be much closer to those to be used for production than before.

Moreover, although the extra planning and simulation have increased the cost at the early stages, the overall cost and the time needed to take the product from concept to production are cut.

Hawtal Whiting, Inc. (Troy, Michigan) which provides a consulting engineering service to the automobile industry, has found that between conventional and concurrent engineering the pattern of spending in the three stages of automobile design differs markedly.

Table 6-1. Ratio of Spending on Design

Stage	Conventional Engineering	Concurrent Engineering
Up to clay model	1	2
Prototype	3	3
Production	5	2

Therefore, although the cost with CE increases in the first stage, the overall cost is likely to be 20 percent less. Planning and design normally account for only 5 percent of the total cost of a new model, and an increase to 10 percent can have a dramatic effect on the quality of the product and the real-life costs. A separate survey showed that Japanese users of the techniques of CE required dramatically fewer work-hours than their Western counterparts using conventional techniques in design.

CE users not only spent the bulk of the design time in the early stages of the project, but they spent half the total time of those relying on conventional engineering. They were thus able to cut design costs or to produce twice as many designs as their competitors. The far fewer hours spent on redesigning the product not only reduces cost but also improves morale of product designers.

Table 6-2. Work-hours in Design

Stage	Conventional Engineering	Concurrent Engineering
Concept	10,000	20,000
Design	20,000	7,000
Redesign	30,000	3,000
Total	60,000	30,000

Failure of Conventional Engineering

The failure of conventional engineering to cope with products can be seen from the minimal number of changes initiated at the early stages of the design. Gradually, the number of changes being made builds up to a crescendo a few months before launch. There is then a reduction in changes until a few weeks after launch, at which point there is a secondary peak that soon tails off.

With CE, the changes are moved back to peak at least 12 months before launch, at which stage most changes can be accommodated easily. Most changes are made at the paperwork stage.

One manufacturer of small ships found that approximately one-third of the design time required for the first ship was required for the second identical ship, principally because so many changes had been made at the end of the program that they had been recorded poorly. Therefore, when the second vessel was built much work had to be repeated from scratch. The Japanese competitor required a negligible amount of time to design the second ship — having only to refer to the original designs and make a few modifications.

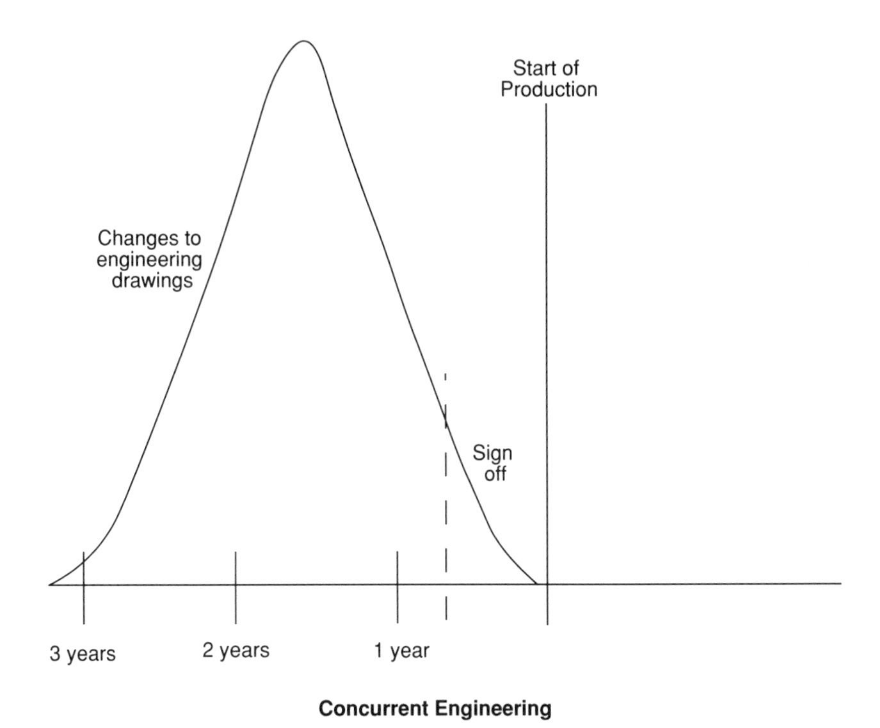

Figure 6-4. With CE, changes can be made earlier so that their cost is negligible.

These examples highlight the benefits of making changes earlier, when the design is still on paper. Then, changes are accomplished routinely before machinery is installed. The existing situation of many changes being made as the product goes into production involves a tremendous number of work-hours from designers, production engineers, manufacturing managers, vendors, and progress chasers — and much of this cost is not logged accurately since it is considered an overhead expense.

The disadvantage suffered by companies using conventional project management and design methods is tremendous and one that will prevent them from prospering in the 1990s. The use of CE is vital not only to succeed in the future, but to survive.

Management Support Vital

However, concurrent engineering is not something that can be pigeonholed under "engineering" while existing management techniques are used. The lead needs to come from the top, with senior managers staking their jobs on the success of CE. They must also give the teams the tools to do the job. First, they must adopt a thorough approach to CE, allowing the task force to develop its skills and giving any training required. Then, they should take a hands-off approach to management, leaving the team to carry the project forward and report back at infrequent intervals.

Some of the bad practices that originated in the days when the boss really was the boss must be eliminated. For example, on a casual walk through the styling studio, the president might tell the modeler to thicken up that rear pillar, or put more of a bulge in the fender, or lower the grill. He has walked through several times before and is convinced that he is right. But that sort of arrogant meddling has no place in a modern project, even if the president does own half the company. If the president or another senior manager really feels that the project is going in the wrong direction, he or she should join the whole team in a roundtable discussion. But it must be with the understanding that the decision belongs to the team.

One of management's main tasks is allocating funds and controlling cash flow. With CE, it must be prepared to provide the necessary funds earlier than is normal in the project, and recognize that money spent on design — not the same as money spent on the product design department — is money well spent. It must be remembered that it is at the design stage that about three-quarters of the total cost of the project is committed, and that any extra cost at that stage will seem like peanuts once the company starts to pay for its tooling.

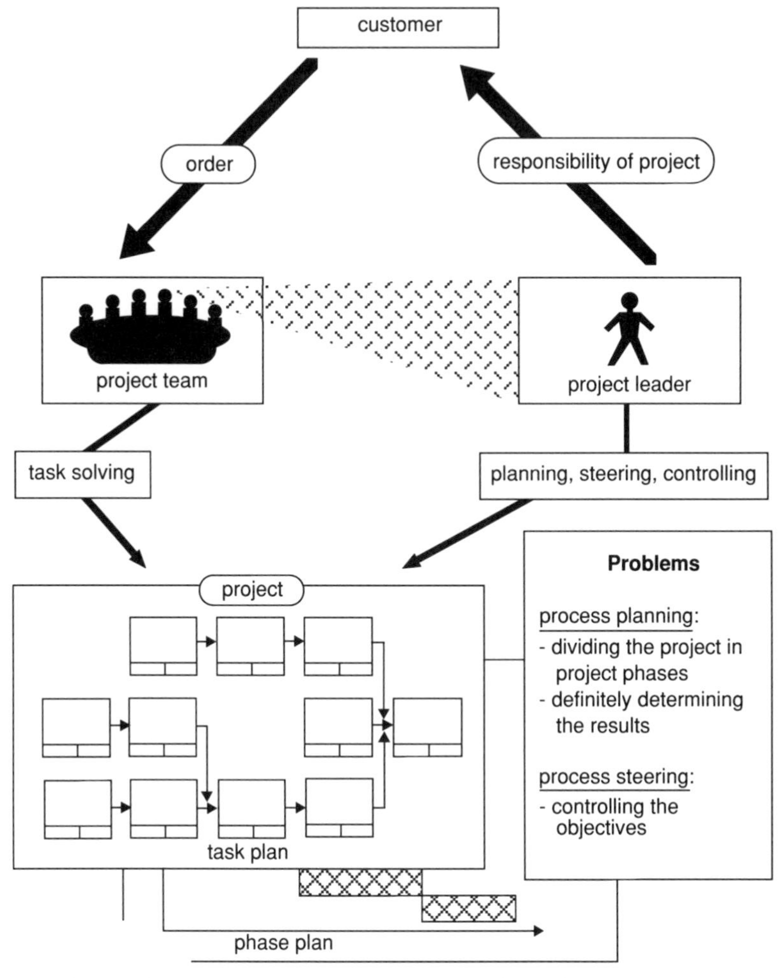

Source: I.W. Eversheim, "Trends and Experience in Using Simultaneous Engineering, " in *Proceedings of the 1st International Conference on Simultaneous Engineering*, London, December 1990, p. 18.

Figure 6-5. Contact between the task force and customers is essential with CE.

CAD equipment and good software are among the essential tools for CE. The team should be capable of simulating as many alternatives as possible on screen, so that by the time metal is cut every practical option has been considered. Equipping the product design office with CAD equipment may represent a substantial investment. However, it can be spread over a period. Investment is inevitable at some stage, even for small companies. When CAD equipment is installed, it must be used to increase productivity. Therefore, management must give the various departments installing CAD equipment time to choose the most efficient hardware and software within the budget. Training for staff is also needed.

Senior management has the most to gain from CE. It achieves

- greater control of product development and production costs
- improved profitability
- greater control of design and manufacturing costs
- enhanced reputation for the company and its products
- higher morale in all departments

In other words, CE is a tool that can dramatically upgrade any manufacturing enterprise and that puts senior managers closer to the lifeblood of their business. It also gives them greater control over the future, as long as they are prepared for change and prepared to let their corporation change to accommodate CE.

CHAPTER SEVEN

Total Quality Control Becomes a Reality

CE fits neatly into TQC
Quality assured at outset by parallel design

Since concurrent engineering results in quality being guaranteed by design, it fits in completely with the concept of total quality control (TQC). Although Japanese companies are generally considered to have developed TQC as an integral part of their business, the basic ideas started in the United States. J. M. Juran took to Japan the idea that quality control was only feasible where it was considered to be an aspect of management, not something to be left to the quality assurance department. W. E. Deming also taught the Japanese about the importance of quality in the whole. So these ideas are as American as mass production; all that is needed is dedication to the ideals.

The basic tenet of TQC, that quality is the responsibility of every employee, is paralleled by the basic tenet of CE — that the quality of design is the concern of all departments. Over-the-fence engineering breeds insularity and rivalry among departments, with a classic response to problems in assembly being to tighten up the tolerances. The real solution is to find out why the product is causing problems and to redesign it. Likewise, when a component is being fitted the wrong way around, the short-term expedient may be to color-code one end — at extra cost — but the real solution would have been to design the component with slightly offset fixings, so that it could not be fitted incorrectly.

Quality by Design

With CE, the control of quality in the true sense is passed back up the line to the task force responsible for design. Good quality is ensured by full use of the talents of task force members and by the techniques used:

- QFD, which specifies the product completely in a matrix
- DFMA, which ensures that it can be made efficiently
- FMEA, which eliminates most unexpected failures
- Taguchi methodology, which leads to a robust design unaffected by variables in the production process
- Manufacturing, which uses required performance levels in assessing production tolerances
- Full documentation of all stages
- SPC (statistical process control), which ensures machines are operating to produce components within tolerances

The task force approach is essential because otherwise, the full skills of the various departments will not be used.

Searching Questions

With both CE and TQC, management needs to be committed to the new methods and prepared to change the company culture to take advantage of the techniques. TQC starts with an appraisal of the business, with some searching questions being asked and answered honestly about the corporation:

- What is the corporation, and what can it achieve?
- What is the corporation doing?
- Can the corporation make changes to itself?

- What are its products?
- Who are its customers?
- Why are they its customers?
- What are the corporation's objectives?

The short answer to the last question is that to prosper the company must retain its existing customers and gain new ones. It will achieve that only if the product and service meet the requirements of the customer, and continues to do so with a level of defects considered acceptable. This brings us back to the customers' requirements and improved quality, key reasons for adopting TQC and CE.

In assessing what the corporation is and what it can do, it is equally important to assess what it has failed to do in the past. This is not an analysis of its lack of size or finance, but of its corporate behavior in crises or in the face of a difficult problem. It is important to be able to assess whether that failing is fundamental or can be eradicated next time around.

What a company is doing may seem obvious, but consider the electronics company that is relying more and more on captive imports to maintain market share, so that hardly any products are made in the United States any more. It can hardly call itself a full-line manufacturer, and if it extrapolates that trend, it will see that in a decade it could be a marketing organization only. The same situation may apply to some automakers in a few years. Is that what a company wants? Will that make it more profitable, more effective? It might well do so, in which case it should continue down that path — but ruthlessly.

Why Customers Buy Your Products

One key question that must be answered is why the customers choose your company, your product. It is usually assumed that some feature of the product is critical, but that is not always the case. When one company investigated that question it discovered the main reason was that it was faster than its competitors in responding to orders. The products had fallen into the category of commodity items as far as the customers were concerned — as is happening with an ever-increasing range of products, from pizzas to semiconductors and trucks — so service was the key, not the product. In that sort of situation, the company needs to reassess its activities ruthlessly. For example, it may be spending the bulk of its advertising and promotion budget on advertising the product, when the customer is not interested in that at all. Also, staff training may be taking a wrong turn.

Quality starts with an appraisal of this sort simply because the effort may otherwise be aimed at the wrong target. Some corporations believe that the appointment of a vice president responsible for quality is sufficient to solve their quality problems, but that approach ignores the all-pervading effect of quality on a company. The person who is ultimately responsible for quality is the chief executive, simply because without adequate quality the corporation's performance will not satisfy stockholders. Therefore, he or she must be committed to the program.

Once a TQC program is instituted, the corporation may benefit from disbanding the quality control department, as both Digital Equipment and Volkswagen did. Quality control is often used as an alternative to good design. In the better corporations, the customer may be assured of a good product by an energetic and thorough quality control department, but at the cost of an unacceptably high level of rejects and with excessive costs for replacement parts.

Right the First Time

Instead, the emphasis must be on getting the product right the first time. For example, Digital Equipment used to need to rework its magnetic disk drives several times to achieve the quality standards it sought. Its engineers were meticulous in not passing off assemblies that did not meet arbitrary standards and in sending them back to be reworked. Analysis showed, however, that with the seven reworks that were typical at that time, the life of the drive was around 10,000 hours. By reducing the number of reworks to three, the life went up to 100,000 hours. If the assembly could be built right first time, the life went far beyond the requirements of the customer.

The solution was to improve quality at the source and to try to avoid reworking wherever practical, even if some arbitrary standards were not reached. This was the last answer that would have been expected, and it would not have been found with orthodox thinking. For this reason, corporations that have fully adopted CE focus on quality, not on tolerances, even though few designers are happy to produce drawings without tolerances on every dimension.

Determine the Real Cost of Quality

Once senior management is aware of the true cost of quality — or lack of it — it will soon recognize the need for both CE and TQC. The failure to approach quality in the correct manner results in hidden costs of substantial levels, as shown in a survey carried out by *Quality* magazine in 1977.

Although the average cost of quality amounts to around 5 percent of sales, it is significant that in almost all cases, failures account for the bulk of the cost. If the number of failures can be cut, costs would be much lower and could easily pay for better quality at the design stage. However, these are the recorded costs rather than the actual costs, which are higher. For example, every time a product fails in service, customers are dissatisfied and are influenced not to repeat their purchase. They are also likely to tell their friends and acquaintances why they will not buy another of those widgets, so one failure may lead to loss of three sales.

Table 7-1. QC Program Quality Costs as a Percentage of Net Sales Billed

	Total quality costs	Breakdown of costs		Failure	
		Prevention	Appraisal	Internal	External
Chemical and allied products	4.81	0.68	1.1	2.03	1.0
Rubber and plastics	14.7	0.40	2.3	9.5	2.5
Primary metal	6.1	0.40	1.47	2.98	1.28
Fabricated metal	5.1	0.51	1.67	1.90	0.77
Machinery	4.4	0.49	1.05	1.76	1.14
Electrical and electronic equipment	5.94	0.90	1.75	2.07	1.21
Transport equipment	3.79	0.34	1.76	1.29	0.50

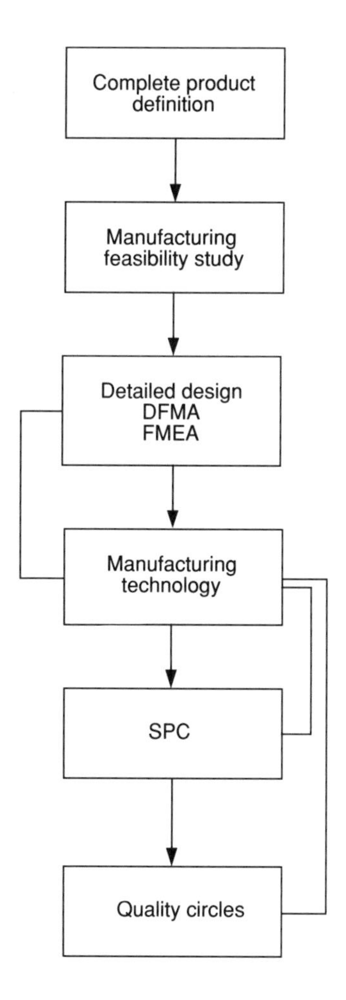

Figure 7-1. Quality begins with engineering and off-line quality control. Quality circles continue this approach, and on-line control is the last stage.

Another survey found that lost opportunities equal the level of failures. In other words, if failures account for 1.25 percent of sales, then the lost opportunities that arise from that failure account for at least another 1.25 percent. Obviously, this sort of data will always indicate a lower figure than the actual level, quite simply because a lost opportunity is not there to be logged. Overall, reducing failures to a level lower than those of competitors results in increased sales.

A much lower level of defects must be a top priority with any chief executive facing Japanese competition, and the combination of TQC and CE is the solution. Once it is recognized that quality is pervasive and that it is everyone's business, efforts can be made to change the culture of the organization to reflect that fact. First, it is necessary to distinguish between quality of design and quality of manufacture, or conformance to the design, as defined by K. Kivenko* and others. Quality of design assures that suitable standards are reached in

- performance
- operability
- reliability
- maintainability
- durability
- economy of manufacture and operation
- safety

Conformance, which is where the traditional quality assurance department comes in, ensures that the product

- meets its specifications
- meets manufacturing standards
- undergoes all processes in manufacture
- is subject to sampling

* K. Kivenko, *Quality Control for Management* (Englewood Cliffs, N.J.: Prentice-Hall, 1984).

- is properly identified
- generates sufficient information for the quality data bank

In adopting TQC, management can opt for a variety of policies. For example, it could extend the principle of QFD by making customer satisfaction the number one priority, and letting quality standards follow that aim. Or, it can adopt a systematic approach as a policy, as a goal in its own right. Although the establishment of quality goals appeals to most managers, it lacks the flexibility and responsiveness of making quality an adjunct to customer satisfaction. Whatever the approach, it is important that quality improvement be continuous and well documented, that it involve all employees, and that it be shown to achieve results soon after the program has started. Managers and other staff should adjust to the need to report on progress in quality on a continuous basis.

It is also important that the performance of the competition be known by all relevant personnel. A system of competitive benchmarks, like that adopted by Xerox, is helpful. In establishing these benchmarks, a procedure is required. For example, the following procedure might be used.

- Plan the benchmarks to be documented.
- Analyze and compare data.
- Develop a set of objectives for performance levels.
- Improve and implement specific actions.
- Review and monitor progress of products against benchmarks.

In putting the aspects of TQC into action with CE, the quality loop defined in ISO 9000 or ANSI/ASQC Q-94 is a useful concept, and is in tune with CE. However, the normal loop, which lacks a hub, lacks a driving force. The CE task force becomes that hub, driving quality through the product and the system.

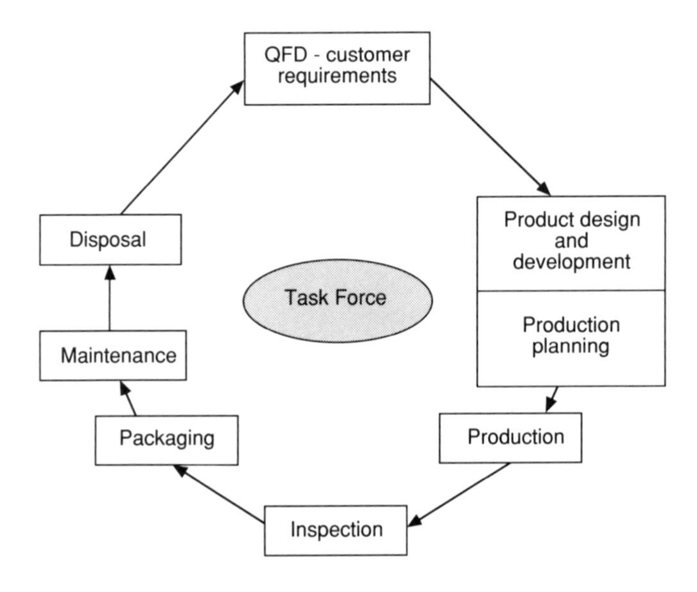

Figure 7-2. A typical quality loop covering the whole life of the product begins, as does CE, with QFD identifying the customers' requirements or desires and ends with the retirement of the product.

Consistent Practices and SPC

Quality in the factory starts with consistency of the processes, which is particularly important in large corporations that make the same product in different plants. Standard processes in all plants will not only ensure consistency in measured criteria but also in unmeasured ones. This can be extremely useful when a new application is found for a product or when some unexpected fault occurs.

SPC (statistical process control) is of course the means of controlling the quality of conformance in the plant. To be effective, SPC is best controlled by the operator under a self-inspection scheme. This requires support from the quality control specialist, through establishment of a workable policy. This is greatly aided by the use of microcomputers and data logging equipment, since the data can be collated automatically, and adjustments simplified.

The basis of SPC is the establishment of the familiar Gaussian bell-shape standard deviation curves for a process. The standard deviation, sigma, is used to indicate the range of the tolerances. The form of the curve is such that with a variation of +/- one sigma 68.27 percent of all products are within tolerance, with a variation of +/- two sigma 95.45 percent, and with +/- three sigma 99.73 percent, which is a typical target for world-class manufacturers.

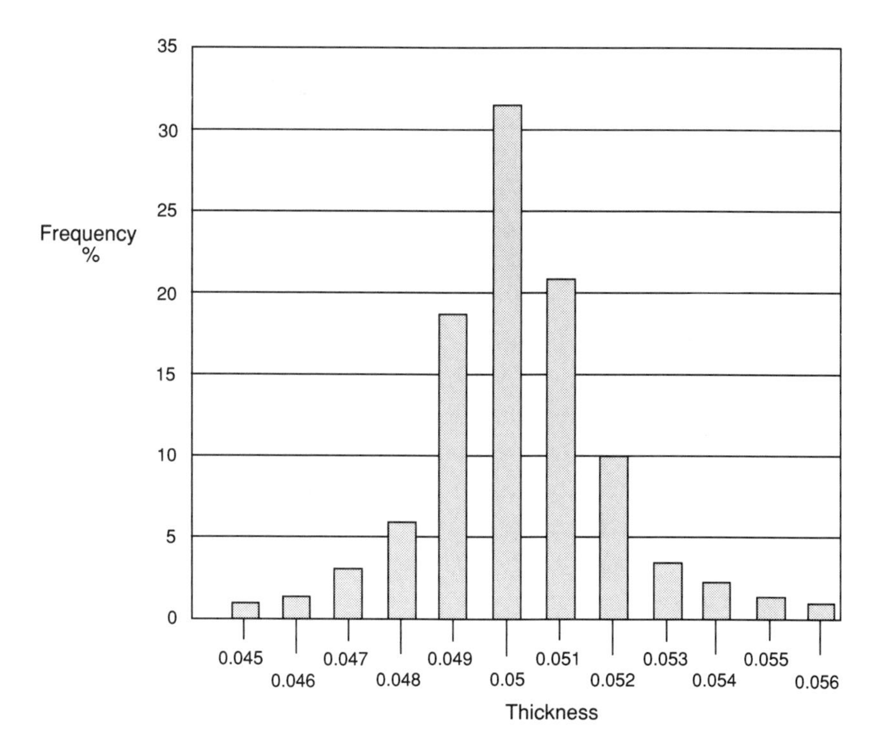

Figure 7-3. Typical distribution of thickness panels

The product can be outside the tolerance either by falling out of the +/- three sigma limits, or despite the close tolerance being maintained, because the mean average shifts from the required level — what Du Pont calls the "on-aim" point.

It is obviously important to be able to distinguish between the two for each process. Otherwise, operators may waste a lot of time adjusting the machine when adjustment is not needed, or adjusting it in the wrong direction. Any time wasted in this way is downtime and usually results in more products being made outside the tolerance limits than if the machine were left alone. So, the waste is serious.

To avoid this waste, personnel need to be given some training and simple "go/no go" charts so that they can work out where the problem lies. At the same time, the data need to be logged so that the statistical data base can be updated.

SPC also has a place in dictating tolerances for manufacture, whether they are selected by product design or by manufacturing: clearly, the tolerance specified by the designer must be one that falls within the normal distribution curve, usually within a six-sigma range. In practice, some provision needs to be made for the range being off-aim. Otherwise, adjustment will be required too frequently, with the possibility of components being produced out of tolerance. SPC data can usefully form a data base in engineering so that designers can find what can be achieved while they are designing components.

Chrysler, which implemented SPC as part of its aim to achieve world-class quality, has integrated it with an on-line factory information and performance feedback system. The results and the performance of the equipment in each process are monitored, and each unacceptable variation is analyzed so that preventive maintenance and machine design may be improved.

This takes us back to the task force, which should ensure that the products are designed so that they can be made easily, and without continuous adjustment of the machines. Of course, in

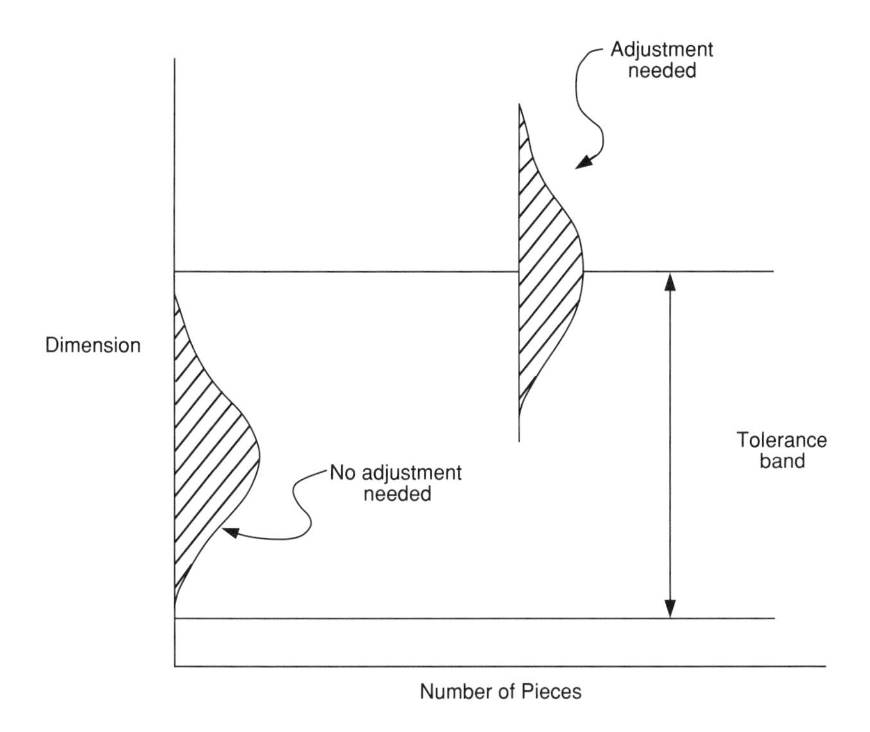

Figure 7-4. When the distribution of product dimensions remains within the tolerance band, machines do not need to be adjusted; but if the whole band of dimensions shifts, then adjustment is needed.

the real world, there are always some machines in a plant that are barely able to produce to the desired tolerances; it is not possible to renew every machine as soon as it becomes a little worn. Special care is needed in monitoring the results from the older machines, so that the ones with wide tolerance bands can be kept in tolerance.

QC Circles

Quality control (QC) circles, which are part of the fabric of Japanese industry, are a worthwhile feature of TQC, giving more responsibility to the employees that want it, as CE gives control to the task force. Normally, the QC circle consists of a group of people that work together in a section or workshop. For practical reasons they are limited to 10 to 20 members, and in Japan people are coerced into joining QC circles in a number of ways — some more subtle than others. In some cases, this coercion is little more than a reminder to employees that any improvements in their sections are good for their company, which is good for them. In others, groups are obliged to submit a certain number of suggestions for improvements each month, and these are tackled by the QC circles. In the United States members join voluntarily, and the meetings are usually held in normal working hours. But that does not diminish the potential of QC circles. Generally, each circle meets for one hour each week, and the leader is normally the section leader or supervisor.

Typically, the section of a workshop that assembles television chassis or automobile instrument facias, or minds a continuous caster in a steel works is likely to form a QC circle. The circle is permitted to investigate any subject, from improved quality of the product to improved working conditions. A management representative gives advice on the operation of the shop, while a member of the production engineering staff gives technical assistance.

In starting up a QC circle, the first requirement is a QC coordinator or champion. This person is responsible for coordinating the programs, ensuring that training material is available, and promoting good communications among circles and between the circles and technical and managerial staff. There should also be a person responsible for getting each circle started. This person, who may be called a facilitator, helps the circle gain some structure in the early days. The facilitator should also encourage circle members to develop their own ideas and methods of operation, and should move out of the circle as soon as he or she can.

The QC circle is a method for increasing the knowledge of the employees as well as improving the quality of the product. It seeks short-term and long-term improvements. Indeed, a survey of users of QC circles found that 73 percent used them to develop the skills of the work force, and only 42 percent gave cost savings as a reason. More significantly, 62 percent of the companies that had abandoned the circle system had initiated them to cut costs. As with CE, the costs come later rather than earlier, and QC circles must be seen as an investment, not as a quick fix. According to one expert, the return on investment in QC circles is from five to eight times the sum invested.

The seven tools of quality control, which should be taught, are

- process flow chart
- check sheets
- histograms
- Pareto analysis
- cause-and-effect analysis, or fishbone diagrams
- scatter diagrams
- control charts

Of these, the Pareto analysis, which ranks faults in order of importance, is one of the most effective in establishing where remedial action is needed. Indeed, it is one of the tools used in designing Taguchi tests. The frequency of faults is ranked, and almost invariably, there are two of three faults that crop up very often. Cause-and-effect analysis is also highly effective.

Another seven techniques, of which the QFD matrix is one, are also used, and at least some of them should be used in training QC leaders. These are the

- affinity diagram
- co-relationship chart
- tree, or distribution diagram

- matrix diagrams
- matrix data analysis
- process decision program chart
- arrow diagram, or program evaluation and review technique (PERT)

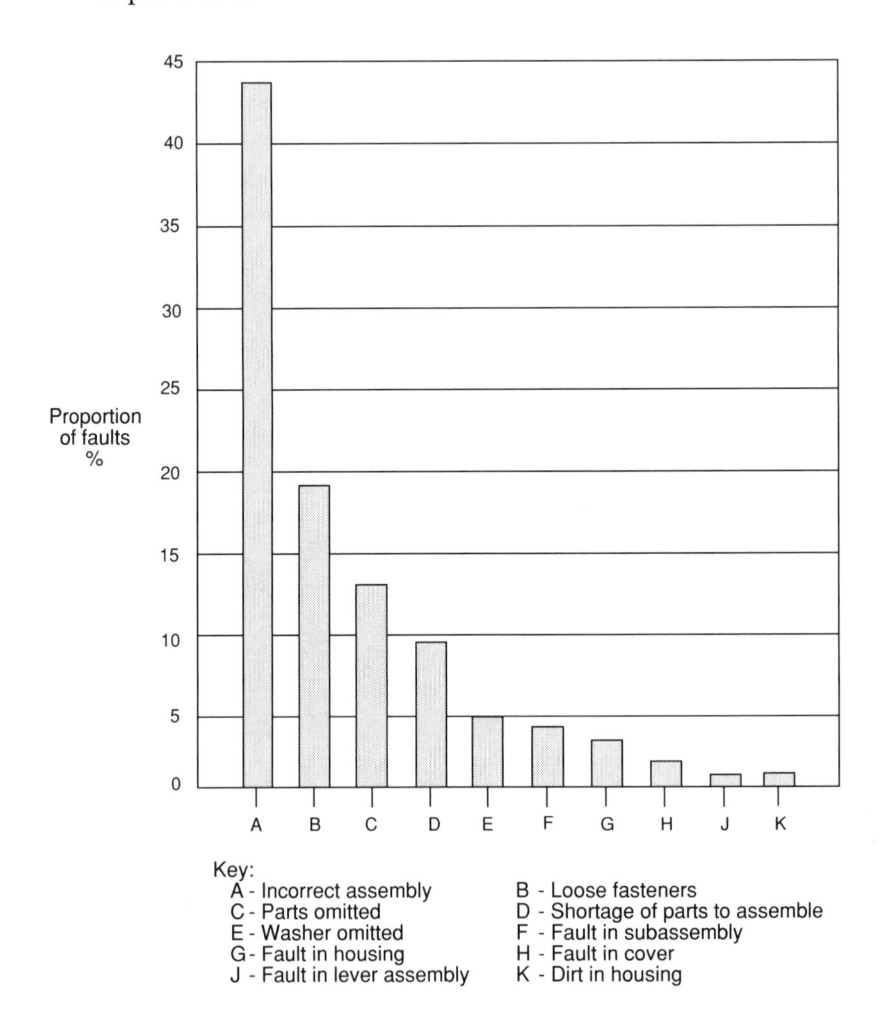

Figure 7-5. The Pareto diagram shows where the problem lies; in this case, faulty assembly is the culprit, not faulty components.

These are intended to stimulate the generation of ideas and to solve specific problems. As in the task force, brainstorming is an essential feature of QC circles. For example, for the affinity diagram, opinions are canvassed from customers or team members, and similar ideas or overlapping concepts are grouped together to stimulate new ideas. This technique is useful where a problem is complex and an urgent solution is not needed.

With the co-relationship chart, a central concept is the starting point. Again, it is used to stimulate ideas, but in this case, the goal is to find logical connections between concepts. Although many different ideas may be brought out, a pattern tends to emerge. Connections between apparently unconnected ideas also become apparent. It is useful where the correct sequence of actions is critical and where the problem is complex.

The tree, or distribution diagram, is used to include all activities required to reach a certain solution or target. Logic is essential, and the procedure starts with the definition of a target, such as to reduce manufacturing costs. The next branches of the tree are to identify critical areas where excessive costs are most serious, and to identify opportunities for automation. At the next branches, the requirements and solutions are broken down again, and the procedure continues until a program has been produced.

The process decision chart is similar to the tree in that a sequence is involved, but the aim here is to include every possible result in a process. It is used to ensure that every eventuality has been considered in an investigation, or the side effect of every possible failure in a mechanism.

The arrow diagram, or PERT diagram, is similar to a Gantt chart, and is used to plan a task. It shows which sections of the work are critical in the elimination of delays and wasted time.

Some managers may doubt that training assembly-line operators in such techniques will produce results. Experience suggests that there are plenty of skills to be tapped, whether it is done this way or in an employee involvement program, like the one Ford adopted before it put the Taurus and Sable into production.

Quality in an organization, like the quality loop, is continuous and pervasive. It starts with the task force adopting QFD to define the customer's wishes; it proceeds through the design into manufacture, where quality is monitored; and it ends up with groups of workers feeding back suggestions to complement the data logged by SPC equipment.

Another important benefit of the use of a multidisciplinary task force is that buck-passing will soon be a thing of the past, so departments will be able to control their own quality. The adversarial situation in which quality control fights production for good quality will be a thing of the past. Therefore, it is reasonable for production to control the mechanics of quality — in other words, to use and be responsible for interpreting results of SPC.

Corporations instituting TQC programs can expect to reduce their real costs of quality by 25 to 30 percent in five years. In adopting TQC, management needs to make its commitment to the changes clear, and to institute a long-term program of three to five years at the start. Everyone needs to be given opportunities and goals in improving quality, and it is important that a project that will bring good results in a short time — three to six months — is started first. The impetus of success is needed to keep the ball rolling.

TQC embraces every aspect of quality, from telephone manners with customers to the waste of copying paper. Therefore, it is easy to involve everyone, and it is important to do so.

Elimination of Waste

The elimination of waste, the linchpin of the Toyota Production System, is an important feature of any quality control system. With CE, waste takes on a wider meaning. Among other things, waste is

- the resources devoted by product engineering to develop products that do not reach production
- the resources required to build unnecessary prototypes
- late changes to a design, which make earlier work obsolete
- the production of stock that is merely afloat
- scrap materials
- rejects
- the cost of the stock of finished products awaiting customers
- failures in the field

With CE, fewer products than formerly will fail to reach production, and because of the greater use of simulation, fewer prototypes will be built. Late changes are also curtailed. Other sources of waste will be reduced because the quality of design and manufacture are improved. However, concurrent engineering will bring the best results in companies that have already made efforts to reduce levels of stock in their plants; have started to pursue waste, including any downtime; and have recognized the need to make changes to be successful in implementing TQC and CE.

Rockwell International, one of the companies that has adopted a companywide strategy to change, defined its plans as essentially beginning and ending with the customer. Its executives decided that so long as they could satisfy external and internal customers, the company's survival was assured — and that achieving that goal would bring about the necessary changes in quality. The company highlighted certain signposts:

- Improving the climate by removing impediments to change

- Defining missions so that everyone knows the goals
- Using the plan-do-check-act (PDCA) cycle to ensure that everything that is planned gets done properly
- Cascading plans from the company's five-year plan, broken down to show what each department should achieve
- Creation of a strategic team to define and oversee the plan
- Adoption of QFD and Taguchi
- Education

Such a program is comprehensive; it requires people to think differently, to demand change instead of resisting it. Rockwell, aware of this problem, put it simply: "The challenge of change; we won't survive if we don't." These words should be remembered by anyone managing a company dedicated to survival and success.

CHAPTER EIGHT

Concurrent Engineering Enhances Design

QFD defines the product in the customer's voice
And results in complete definition
Fewer late changes with improved quality

Despite the fact 60 to 80 percent of the total cost of a product — materials, labor, capital investment, and ancillary costs — are committed at the design stage, many senior managers consider design a necessary evil, a costly nonproductive unit. Yet the cost of design engineering is unlikely to exceed 5 percent of the total budget of a project. If it were to increase to 10 percent that could be a benefit — so long as the result was better design and quality, and easier manufacture.

The Ford Motor Company found that although the cost of materials in a product is 10 times that of design, it affects only 20 percent of the cost of an automobile. By contrast, it found that design influenced 70 percent of the cost — figures that indicate where companies should increase their investment.

Evidence that an overrun in the cost of engineering new products is among the least of management's problems comes in a survey carried out by McKinsey and Company. It found that the biggest killer of profits is bringing a product late to market.

Table 8-2 shows that the most important priority with a new product is for it to reach the market on time, and that lateness to market cuts profits more than does exceeding the desired

Table 8-1. Influence of Design Activity on Total Costs

	Relative cost at time of manufacture (%)	Influence on overall cost (%)
Overheads	30	5
Design	5	70
Labor	15	5
Materials	50	20

Source: Ford Motor Co.

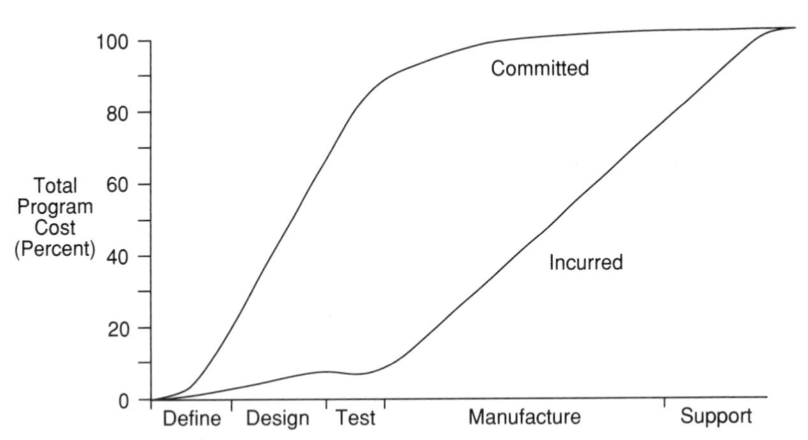

Source: D. Grant, "Simultaneous Engineering in Aero Gas Turbine Design and Manufacture," *Proceedings of the 1st International Conference on Simultaneous Engineering*, London, December 1990, p. 116.

Figure 8-1. Although the bulk of the cost of a program is committed at an early stage, most of the cost is incurred in tooling up for manufacture.

production cost. Even if the cost of product development is 50 percent over target — a situation that could lead to the chief engineer being fired — the effect on profits is marginal.

Table 8-2. Reductions in Profit Caused by Different Problems

Product six months late to market	33%
Product cost 9% too high	22%
Cost of product development 50% over target	3.5%

Source: McKinsey and Co.

Of course, with the cost of product development accounting for about 5 percent of the total cost of a product, it is easy to see why an overrun is not a serious problem — so long as the product benefits. Poor quality is similar to lateness to market, because sales will be below target. CE not only improves the chances of bringing the product to market on time, but it also results in improved quality.

By contrast, over-the-fence engineering shackles the designers even in well-run companies, and results in excessive cost. Right at the beginning of the concept stage, conventional engineering starts to take the project in the wrong direction. Usually, concept work is done under great secrecy in restricted areas of the R&D department. Unless it is essential, no one outside the group is involved, not even the product designers who will eventually draw up the detailed design, the manufacturing engineers, or the vendors of critical assemblies of raw materials such as sheet steel or castings.

For example, in ascertaining how much space a new type of instrument might require, outside help may be sought — but the vendor would receive information on a "need-to-know" basis. Yet the engineer undertaking the briefing does not know precisely what information the vendor needs, and may not be aware that a little more information could lead to an improved product. He or she may be giving the vendor a cube of a certain size as the envelope, when in fact a column the depth of the cube could extend from the cube without affecting the installation. Unless the vendor's designer has direct contact with the company's design office — often there is no direct contact at all — he or she is unlikely to design an assembly that exceeds the specified dimensions. Yet that limit, unnecessary in one plane, may increase the cost of the assembly by 30 percent. Poor communication also leads to increased cost of castings and forgings.

Major Drawback of Over-the-Fence

A major drawback with conventional engineering is that normally the concept will be taken through to the design stage without any detailed assessment of how easily it can be produced — and unless a manufacturing feasibility study is detailed, it is unlikely to reveal the problems. In many cases, the definition of the design will lack detail as well; the effort is aimed at obtaining the best design for its purpose on the assumption that changes to suit manufacture or the customer will be minor and can be handled later. Ironically, many product engineers are extremely reluctant to change the design at this stage, but a year or so later, when the design is nearly finalized, they are more responsive to requests for change — the exact opposite of what is required.

In some cases, experts in manufacture will be brought in at the design stage — but only to make comments on the suitability of the existing design for manufacture, and to suggest minimal changes. The design engineer is thinking in terms of wall thicknesses, draft angles and boss sizes on castings, and bend radii on pipes. He or she is not asking whether the casting could be made more easily as two pieces or whether a housing consisting of two castings would be better for the foundry than a housing consisting of one or three.

This situation results from the way in which engineering companies operate, coupled with the responsibilities of departments. It happens however skilled the engineers may be, and the attempts of a few forward-looking thinkers to break down the walls are generally resisted by those around them. When a task force is first set up, some design engineers feel threatened. Careful management is needed to allow them to concentrate on their skills, leaving other experts to concentrate on theirs. Care is needed in defusing any attempts by middle managers to play company politics as well.

Inherent in the system of over-the-fence engineering are many weaknesses. Products are taken from the concept stage toward prototypes with

- insufficient definition of the product
- no design for manufacture and assembly (DFMA) studies being undertaken
- no clear guidelines on how they will develop in detail before production
- ballpark figures on costs only
- enormous potential for late changes

CE Solves the Problems

Because it is based on the task force concept, in which all members are equal, and because it embraces certain techniques, CE overcomes these problems. The equality of members is important: should the production engineer make a suggestion that appears to change the form of the product drastically, the design engineers cannot be allowed to say simply that it will not meet the design criteria; or worse, to tell the production engineers to mind their own business and let the designers decide how the product should be designed.

Such jealousy of one's special responsibilities runs against the business interests of the corporation. The loyalty should be to the corporation and its product first. The function or skills are there to be exploited, and that means the complete team needs to discuss the matter from every angle. Then the necessary schemes, costs, and so on need to be undertaken. Clearly, the design engineer is responsible for functional design, but since the task force is a team effort, anyone can make suggestions in any area.

Even though they may not have adopted CE, many large corporations have adopted specific programs for product development, with definite points at which responsibility is signed off. Such a program is a necessity with CE, for several reasons. For example, maintenance and the retirement of the product are considered part of the process, because one of the tenets of CE is that the costs must be judged over the whole life of the product. Then, one of the aims is to formalize the activities so that people can be managed by results. Without a formal procedure, in stages, there are too many uncertainties in measuring performance.

Stages in Product Development

When it conceived its product delivery process (PDP), Xerox identified eight stages in the life of a developing product:

- Preconcept: an idea for a new product is turned into an outline specification
- Concept: the specification is matched against requirements and is defined
- Design: the engineers design the product to match the specification provided
- Demonstration: the building of prototypes to ascertain correct functioning, followed by pilot production to remove any production problems
- Production: manufacture starts
- Launch: marketing and sales start
- Maintenance: service requirements throughout its life
- Retirement: replacement by a new model

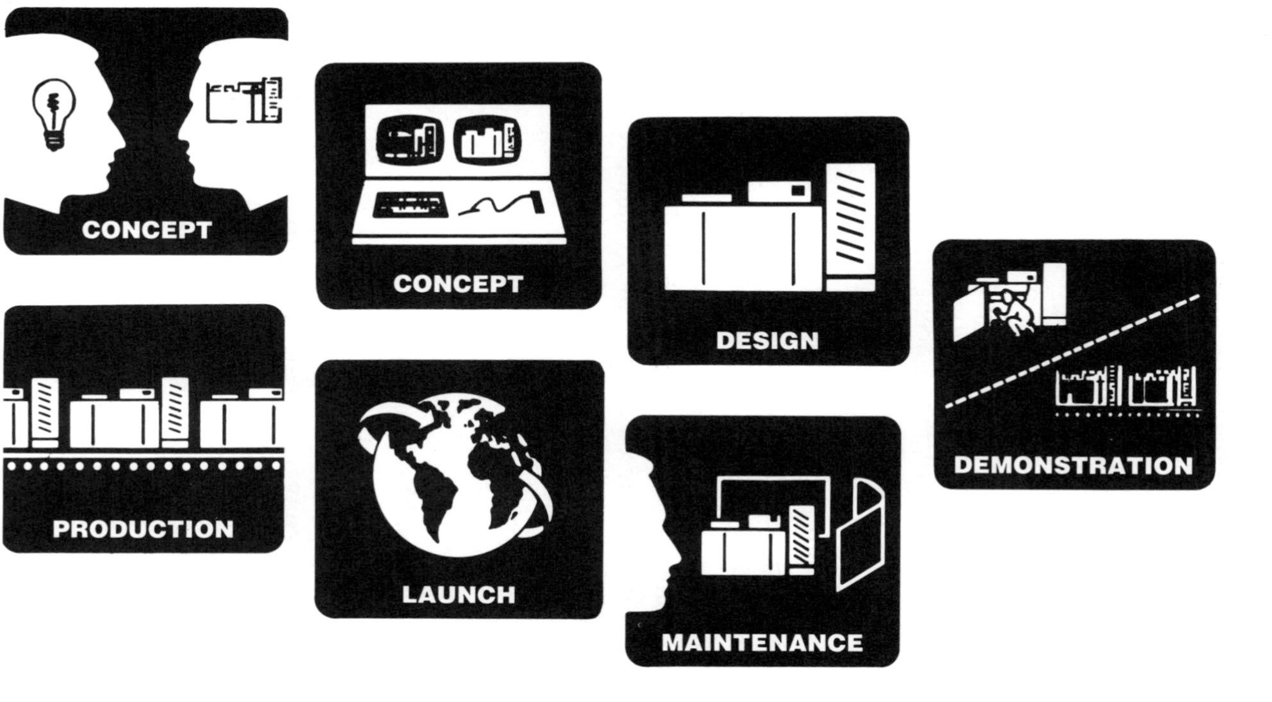

Source: Rank Xerox Limited

Figure 8-2. Xerox defines seven stages in the life of its products and the product delivery process.

Other approaches divide the program into four stages:

- Concept
- Design and development
- Design validation
- Production process development

Whatever approach is used, it is important that the end of each stage has a definite timetable and specific performance requirements. For example, the concept stage must include a feasibility study on production techniques, and the study should include quantified results — not a statement from the chief production engineer that the product can be made. CE is all about quantifying problems and solutions.

At the concept stage, it is often difficult to assess which is the better of a number of alternatives. One approach is concept selection, sometimes called the Pugh method after its originator Stuart Pugh of Strathclyde University in Scotland. It evaluates products against chosen parameters and is designed to spark innovation; this replaces arguments over whose concept is best.

Criteria, such as speed of operation, weight, initial cost, and maintenance intervals are listed in the left-hand column. Alternative concepts are listed across the top of the grid, and each is rated +, - or =. The best concept is not adopted at that point, but efforts are made to improve the weakest points of that concept to the best level of any other concept. Thus, the alternative concepts converge to an optimum. This technique, which is part of Pugh's total design concept, has been adopted by a number of American companies, including AC Rochester (Rochester, New York), and was used in the development of GM's Saturn project.

Also at this stage it is important to take into account the timing of the introduction of the product and the timing of the commercialization of any new production techniques. There are many products designed for a brand-new process that promptly fails to live up to its promise, so the product has to be redesigned and is late to market. In some cases, the performance suffers drastically as a result. The timing for the introduction of new processes needs to be conservative, but new processes should not be ignored because risk is involved; rather, the risk should be quantified and contingency plans made.

Quality Function Deployment (QFD)

Once a task force has been set up to develop a new product, the first task is to define the product. But how is the product defined precisely at the concept stage? Quality function deployment (QFD) is the best technique available.

QFD ensures that the product meets the customers' requirements when it goes into production just as well as at the concept stage. It also results in the design being specified more fully earlier than is normal, thus allowing changes to be made earlier. It eliminates the possibility of misjudging a certain feature as a customer's requirement when it is not, and later on is deemed to be the critical factor in the design — simply because it has been passed down the line as an important feature.

QFD was developed in the late 1960s by two Japanese professors, and used first in the shipyard of Mitsubishi Heavy Industries at Kobe in the early 1970s. It has since become used widely by Japanese manufacturers and is one reason why they are able to judge customers' requirements well. In a survey of 135 large Japanese manufacturing companies in 1986, 50 percent were found to be using QFD.

Professor Yoji Akao, one of the originators of QFD, defined it as the deployment of quality, technology, cost, and reliability. There are two aspects to this:

- Quality deployment, which converts customers' requirements into substitute characteristics, and which establishes design quality of a final product. The relationships

are deployed systematically, starting with the quality of each functional component, and taking these through to the quality of each component and process.
- Quality function, defined as the area of responsibility in manufacturing companies through which fitness for use — quality — is achieved.

Significantly, quality is seen as the essence of design and manufacture. The aim is to ensure that quality is obtained through design, not afterwards. This is analogous to the American approach, in which quality is divided into quality of design and quality of conformance, referred to in Chapter 12.

Quantified Problem of Rust

An early example of the use of QFD was when Toyota was suffering, like other Japanese automakers, from excessive rust in its cars. Toyota engineers met customers who were complaining about the rust, and translated the customers' complaints into engineering technology. Toyota also compared the rusting of its cars with the competition, and then decided how far it should go in matching the customers' requirements. The result was not just the introduction of antirust measures but the knowledge that the performance level they had chosen would satisfy the bulk of their customers. It was a truly cost-effective solution that could be introduced with confidence.

Too often when there is a serious customer complaint, engineers indulge in overkill. Because everyone in the company knows there is a serious problem, the cost is not called into question; the priority is to kill that problem stone-dead. With too many urgent complaints by customers to be rectified in this manner, the cost of production soon renders the product uncompetitive. The combination of CE and QFD prevents occurrence of this train of events.

QFD is not just about the requirements of customers when they play their radios or watch TV. It is also about the product's servicing. If a product is difficult to service, specialists will not want to handle it, or will charge highly to do so. In the end, that cost or difficulty becomes a cause of resistance among customers. With QFD, the aim is to prevent such a situation from arising.

For example, when Northrop applied CE to the development of the advanced tactical fighter plane, it found that a key requirement was that the plane should be able to land, go through a maintenance procedure, and take off again in 15 minutes. So, it was decided that one of the customers' voices that should be heard was that of the technical sergeant who maintains the airplane. Northrop personnel spent some time at an Air Force logistics center to find out just what features the technical sergeants wanted. As a result, the plane should meet that part of the specification with ease; at the time of writing, the plane is still a long way from service.

Customers Add Voice to a Specification

The whole point of QFD is that it gives the customer's voice due importance. It is an ideal way of turning the vague preference of the customer into an engineering specification. To be specific, QFD turns customer comments like "a quiet engine" in a car into more specific terms such as

- an engine that is quiet at 70 mph
- an absence of vibrations through the toeboard at idle
- quiet enough to idle in the drive at 2:00 a.m. without waking the neighbors

These terms, sometimes called "wants," are then turned into engineering specifications — the "how" list, such as actual noise and vibrations levels.

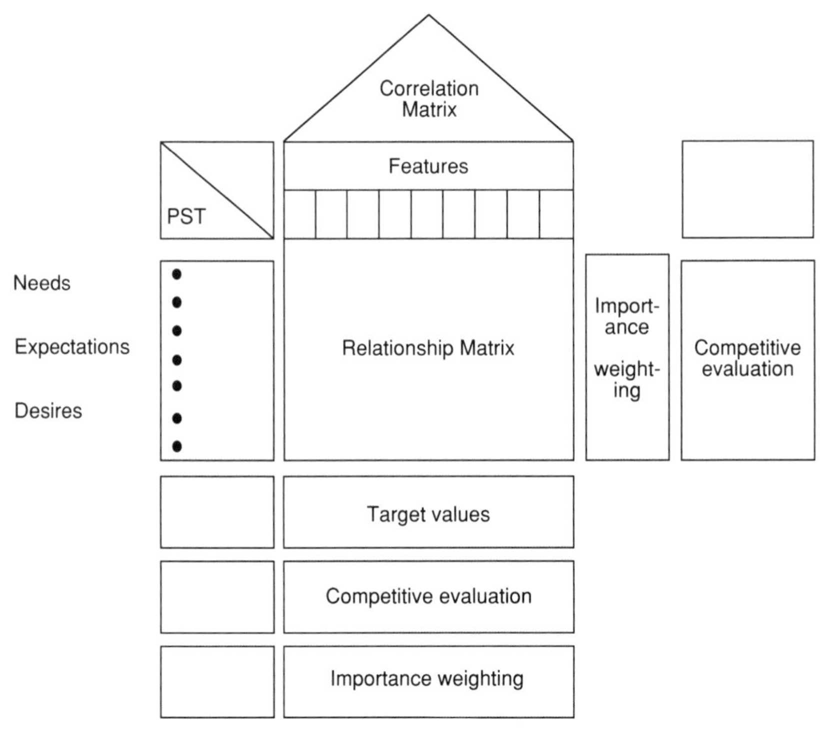

Source: Rank Xerox Limited

Figure 8-3. Xerox's house of quality is a specialized approach to quality function deployment.

With conventional engineering, marketing managers, product planners, and design engineers tend to imagine what the customers' requirements are — after all, they reason, they use the product, so they should know. Although that is true, they all have their own prejudices, which result largely from their professional involvement with the product. This is quite unacceptable. Essential principles of QFD are that actual customers are asked what they want and that their answers are preserved in their words. This exercise can take the form of market research surveys and clinics in which people in the planned market sector are shown concepts. Normally, the "customers" will be owners of the existing products, but when a product is intended specifically to take customers from a certain competitor, then owners of that competitor's product should provide the customers' voice.

The view of customers inside the company — manufacturing, sales, and service engineers — must also be considered. All these people will benefit from a good design, and will act with the indignation of customers to any problems with the product while it is in their sphere of operations. This, of course, is why the manufacturing engineers and marketing people are in the task force in the first place.

QFD Chooses the Right Product

Clearly, the process of identifying the customers' requirements becomes a major operation, and one that merits maximum priority from management. A good example of the influence of the voice of the customer occurred in the development of the high-performance version of the first Honda City. The engineers at Honda R&D, the company that acts as the product engineering department for Honda, were convinced that an engine with four valves per cylinder was the optimum design in terms of performance, economy, and whole life costs.

However, marketing found that potential customers for small high-performance hatchback cars in Japan at that time wanted turbocharged engines. Honda engineers pointed out that for the turbocharging to be successful, electronically controlled fuel injection would be needed, so the cost would be greater than for a four-valve head with carburetors, which could give equal performance. Once customers indicated that they wanted turbo power, the engineers shrugged their shoulders and developed an excellent turbocharged engine for the City.

The car proved highly successful, but four years later, when Honda was designing a second range of high-performance engines, the customers' demands (voice) had changed; now they would be happy with a four-valve engine. So that is what Honda, and all its Japanese competitors, started to produce. It should be added that Honda did not at first achieve the correct balance in market demand. In Japan, the company had expected that most customers would want the 2.0 liter four-valve engine for the new Accord in 1985. When this was not the case, it had to add a 2.0 liter unit with three valves per cylinder to satisfy demand.

Thanks to the task force approach, Honda was able to rectify this failure to gauge the proportions of the different segments of the market. Is that a criticism of QFD? No; nothing is perfect all the time, but QFD comes closer to the mark than anything else and puts the task force in the right frame of mind to adjust to changes in the market.

Many engineering departments faced with the situation Honda faced at the concept stage for the City would have designed the engine they thought best, and lost the chance of a good market share for relatively little investment in new plant and tooling. Instead, they would have designed an engine requiring new investment in plant. Since they had failed to read the customers' requirements correctly, sales would be poor. By the time the engine of the type they had developed had become popular, their model would still not have broken even. Little wonder then that management would be reluctant to sanction an extended high-performance engine project.

So the failure to discover what the customer wanted would lead to the loss of two opportunities:

- Good sales for a relatively small investment
- Management go-ahead for a second high-performance engine project

Another problem with a conventional approach to marketing is that the detailed specification of a product — for example, determination of whether treble and bass controls are incorporated in a cheap audio system or whether cruise control is standard on an automobile — may be at odds with what the customers actually want. When Digital Equipment started to question its user base about the features it required on minicomputers, it found that many did not want things that the company had thought important. On the latest models, which were specified on the lines of what the customers actually wanted, there are now 20 percent fewer features, at a considerable saving in cost and gain in reliability.

QFD can therefore push an idea that is ahead of its time back to where it belongs, or show what features are not needed. In the meantime, it also leads to a product that the market wants, and it positions the designers close to the marketplace.

Matrix Specification

With the initial stages of QFD performed, the company understands what the customer wants in detail. Next, the customer's requirements — in the customer's words — are converted into quality characteristics. A quality planning matrix is used to define the characteristics of critical parts. Then the company plans the process, identifying the critical points, to achieve the desired characteristics. Finally, it plans production to ensure that the original requirements are met.

In drawing up the quality planning matrix, customer wants are first rated according to the importance attached to them by customers. So important is this procedure that great care is needed in wording and asking the questions; any loading of the questions will obviously distort the answers, and this is one reason why all the members of the task force need to be involved. The results produce a wants list that forms the left-hand column of the matrix.

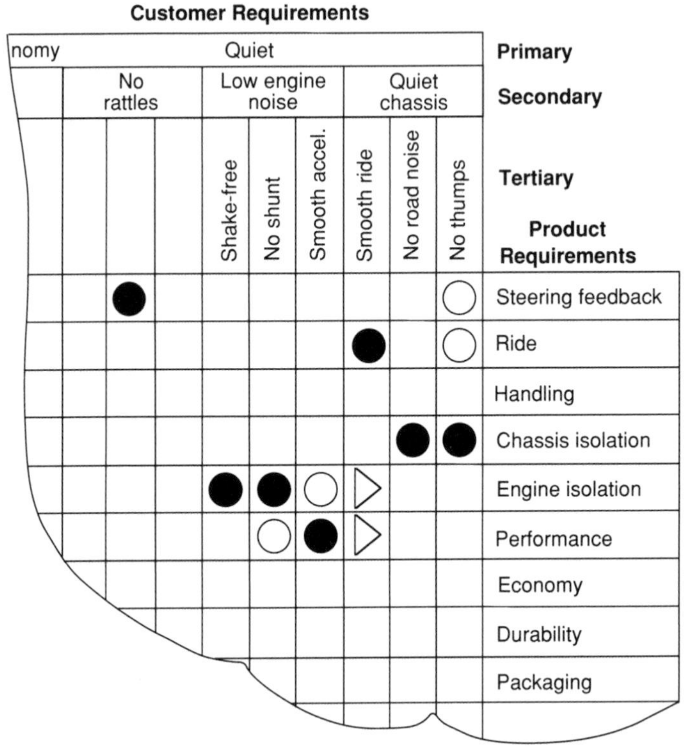

Customer Requirements

nomy
Quiet
No rattles
Low engine noise
Quiet chassis
Shake-free
No shunt
Smooth accel.
Smooth ride
No road noise
No thumps
Primary
Secondary
Tertiary
Product Requirements
Steering feedback
Ride
Handling
Chassis isolation
Engine isolation
Performance
Economy
Durability
Packaging

Source: Rover Group, Ltd.

Figure 8-4. Customer requirements are the basis of the QFD matrix.

Subsequently, the wants are compared with the engineering requirements, which are listed across the top of the matrix. The correlations between them are rated from "strong" to "weak." For example, engine mount design has a very strong relationship with the level of vibrations in the toeboard, but is likely to have a weaker relationship with noise level in the passenger compartment.

It is also important to establish the performance of close competition, which is usually recorded in the column to the right of the matrix. The views of the customer, the levels of competition, and the engineering criteria form the basis of the QFD matrix. Then, an engineering specification can be drawn up, and process sheets produced. Throughout, a matrix of these and other requirements is included.

Therefore, everyone involved knows the ultimate aim of the product in terms of the customers' voice. The production engineer is not wondering whether it is necessary to maintain the specified tolerance on crankshaft balance; he or she can see what level of vibrations is acceptable, and together with the design

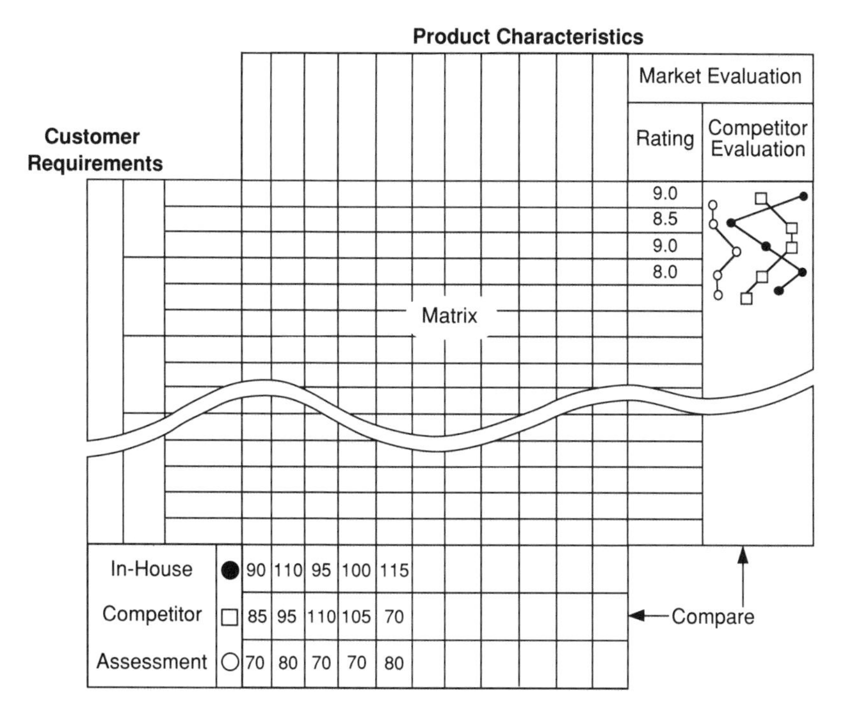

Source: Rover Group, Ltd.

Figure 8-5. With the competitor's products analyzed, the QFD matrix can be used by all departments.

engineer can organize tests to find how that level can be obtained.

The inclusion of competitors' products is also important because everyone in the project team knows what has to be done to match the opposition. In many companies, such data are hidden away in the competitive products department, where few can see them.

Another problem with conventional engineering is that it is common for a forceful personality to insist that a competitor be matched in one respect for which there is no actual customer requirement. Consequently, cost is incurred unnecessarily, while more urgent factors are ignored. With QFD, the customers' voice takes precedence over domineering personalities.

Judgment Needed

Judgment is still needed throughout, and here the task force will use the information concerning the customers' voice and competitors' performance to establish actual targets. For example, the company may have the aim of moving up-market. Unlike some competitors relying on their own judgment, these engineers will know from the specification in QFD that the addition of some brightwork or a few strips of wood veneer on the facia combined with higher prices will not succeed.

They will see that standards of refinement and defect levels of products in the higher class must be matched. During the design specification stage it may be decided that in order to position the product well into the higher class, the level of refinement, ergonomics, and tactile impressions must equal the best of the competitor's performance. It will follow that it will be impractical to match the best in some other aspects, such as low cost or weight.

In another case, price may be paramount, so a level of refinement closer to the middle of the range of the competitors may be considered appropriate. It can be seen that QFD is the essential starting block in CE, and will ensure that the design is specified properly, and that the specification can be obtained in production.

Moreover, should customer resistance be encountered, then QFD is the way to find out why. Because of the need for accurate data on the customers' views and for an understanding of those views by product design, marketing, manufacture, and finance, QFD is only completely successful when used in a CE environment.

CHAPTER NINE

Easier Production, Fewer Failures

DFMA (design for manufacture and assembly) a vital tool
Cuts parts count by 20 to 60 percent, parts costs by 30 to 60 percent
FMEA (failure modes and effects analysis) to reduce failures

For many years, the automobile manufacturers have been preaching the importance of designing components and assemblies that could be made easily. To this end, they paid great care to the details of designs and to materials specifications. The emphasis was on feasibility and machinability of materials. Indeed, feasibility engineers worked in the product design departments to advise on production techniques, and their job was to eliminate the big snags, the components that could not be made. Product designers made efforts to incorporate the advice, but most of the time they found that major changes impaired the design — or so they thought.

Until the automakers started using CE, attempts to change designs radically were largely unsuccessful. The reason was simple enough: even where great efforts were made to alter the design to simplify manufacture, the recommended changes were made too late, and in any case were usually too minor to make significant cost savings.

A classic example occurred when Xerox applied DFMA techniques to a latch on one of its machines. Significant improvements were made in redesigning the latch, as shown in the Table 9-1.

Table 9-1. Xerox Latch Design Improvements

	Old Design	Design with DFA
Assembly time	7 min.	1.5 min.
Assembly cost	$2.7	$0.6
Parts cost	$9.8	$7.4
Total cost	$12.5	$8.0

The gains were substantial. However, at the time, Xerox was not using task forces in product development. Although the component was redesigned by product design, the procedure took too long; by the time the work had been completed, production was too close for the redesigned product to be used. Clearly, DFMA alone is not a solution; it needs a CE environment to be effective.

DFMA and CE

This is not surprising because in over-the-fence engineering, DFMA becomes one more excuse for acrimony between product design and production engineering. Indeed, it was the need for the use of a multidisciplinary team for DFMA that led Ford to adopt concurrent engineering.

Once the task force approach is adopted, the implementation of DFMA follows naturally. All the team is involved in the design, so simplified production techniques can be considered before the designer starts work. Second, since the team is involved at all stages, it can suggest changes before it is too late.

Third, the presence of an expert on purchasing results in instant answers on the cost of different materials and vendors, and on whether to make or buy the product.

DFMA is not a tool to be added at the end of a project. Like CE, it is a way of life that must be part of the project from start to finish.

There are various techniques and sound practices for DFMA. The general aim is that the components can be picked up easily by human or robot, and that they are assembled by stacking. One component can be shaped to act as a guide for assembly of the next one, especially if there is a possibility of the part being installed improperly.

Stacking with Guides

The components should be designed so that none need be assembled from an oblique angle. The lot of the production engineer is improved if different faces of components are identified as

- functional faces, such as the faces of gear teeth
- connection faces, such as the bore of the hub
- free faces, which do not have a direct function

This approach enables engineers to give appropriate attention to each and prevents unnecessarily tight tolerances from being applied to faces that do not need them.

In some cases, a fabricated metal housing consisting of several pieces of different shapes and thicknesses can be best replaced by a cast housing; in others, a casting can be replaced by a stamped steel plate, with plunged holes tapped to retain fasteners. Posts and pegs can be projection welded or riveted in position. Then, the cost of assemblies can be reduced with the casting or molding in place of pegs or spindles. The use of an adhesive to retain a bearing in a recess can eliminate the need for a close tolerance fit. Plastic moldings can be designed with integral hinges or snap connectors. These are just a few of the principles adopted by specialists in design for manufacture.

In some cases, the change seems to contradict another recommendation. For this reason, a number of specialists have produced systems to rate designs for manufacture. Hitachi, the Japanese electronics giant, developed a system that it exploited in improving the design of its videocassette recorders and audio decks for automated assembly. GE of the United States took up this technique and added criteria for parts count, creating a version that GM used for some time. In Britain, Hull University developed a system for use on a personal computer, and this is marketed by Lucas Industries, which uses it in-house.

The Industrial Technology Institute (ITI) (Ann Arbor, Michigan) has developed a system it calls the integrated product/process design method, in which there are five steps. It defines alternative product and process concepts. Where manufacturing does not pose any limits, the design is product driven; where new manufacturing techniques are used, the process-driven approach is adopted. In addition to this conceptual approach, ITI also sets producibility goals according to the method of assembly.

General rules for DFMA, recommended by a variety of specialists, include the following:

- Use the minimum number of parts.
- Make designs modular.
- Minimize variation in parts.
- Where variety is inevitable, arrange for the unique parts to be assembled last.
- Design parts to be multifunctional.
- Design for ease of fabrication.
- Avoid separate fasteners.

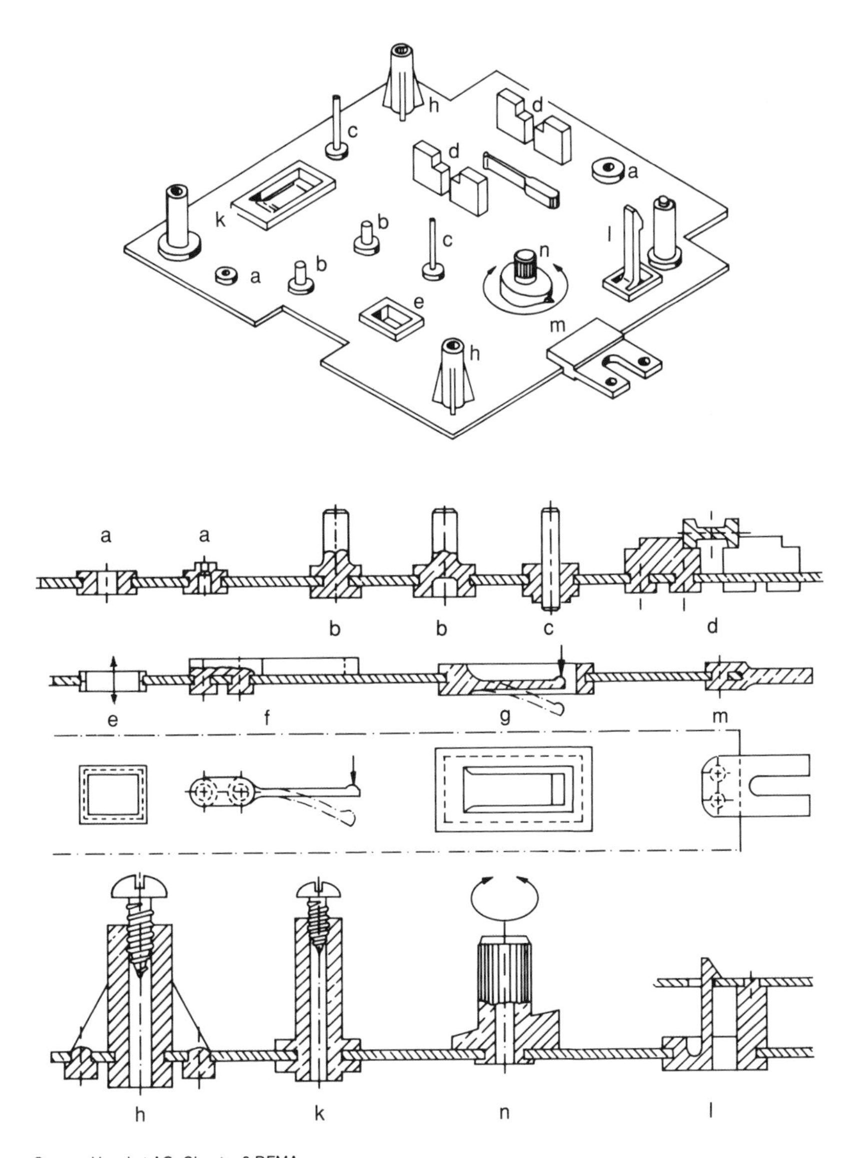

Source: Hoechst AG, Chapter 9 DFMA

Figure 9-1. Ideas for simpler assembly: by molding between components and a sheet metal plate, a chassis can be produced simply.

- Minimize assembly directions, as each operation should follow the previous one.
- Design for ease of assembly.
- Minimize handling.
- Eliminate or simplify adjustment.
- Avoid components made from flexible materials.

The most comprehensive approach is the system developed by Drs. Geoffrey Boothroyd and Peter Dewhurst at Rhode Island University. Manuals and PC software are available from their company, Boothroyd and Dewhurst, Inc., which numbers Black & Decker, Digital Equipment, Ford, General Motors, IBM, NCR, Rockwell International, and Texas Instruments among users of its software packages.

The principle is that the system analyzes the assembly in a number of ways and assesses it for ease of assembly or manufacture — there are different packages for different uses. As a general rule, a good assembly is one that has few components, since it will take less time to assemble, and is less likely to be assembled incorrectly or to fail in service.

Fewer Components, Less Work-in-Process

Management has much to gain from DFMA because its benefits extend far beyond lower assembly costs. First, since fewer components are required, the assembly is inherently more reliable, and easier to repair should a fault develop. Production managers will find that work-in-process can be reduced, as demonstrated by Motorola, which attributes a reduction of 40 percent in work-in-process to the use of DFMA. Therefore, goods-receiving warehouses are smaller, as are those for replacement parts. Overheads are reduced because there are fewer drawings to be stored and maintained, fewer replacement parts to be held and shipped, and less documentation.

Taking into account all these costs, NCR calculated that the true cost of every fastener in its cash registers over the whole life of the machine is $12,500 — just for one bolt. The use of DFMA almost invariably results in a number of fasteners being eliminated from a small assembly, with significant long-term savings.

Analysis of Assembly of Product

Boothroyd and Dewhurst's DFMA analyzes the structure of a product and estimates its cost and assembly time. Each component is assessed on a scale from zero to nine for the ease with which it is grasped and oriented; and for its thickness and size. Standard times are determined from these values. This allows a total time to be calculated for the assembly. Next, an attempt is made to simplify the assembly by eliminating parts. Each part is examined and three questions asked:

- Does it move relative to other parts?
- Does its material need to be different from that of the others?
- Does it need to be a separate component, or could it be combined with another without compromise to functionality?

If a separate component is not required — and a fastener is never required on this basis — then efforts should be made to eliminate it. Also, if there is no relative movement between two components, as in the case of a top cover for a housing, the need for a separate part is queried.

The important feature of the program is that it does not attempt to redesign components; that is left to the designer. It merely rates the products on how easy they are to assemble and prompts the designer on which parts can be combined or eliminated. There are many cases where a housing consists of three castings because that is how it has always been done. DFMA prompts the designer to ask questions that may have been forgotten. The result is almost invariably fewer parts and shorter assembly cycles. Ford, one of the big exponents of DFMA, has found that it reduces parts count overall by about 33 percent.

DFMA is also part and parcel of Chrysler's approach to CE. Typical of the gains Chrysler made was in an instrument cluster, where a two-day DFMA analysis by the team resulted in the parts count being reduced from 60 to 25 parts, with considerable benefits. Delco Remy considers DFMA one of the formal structured processes that improve CE.

In practice, many of the old methods of designing for manufacture — such as folding the sheet metal in only one plane — are shown to be inappropriate by DFMA. For example, it may cost far less to replace three simple sheet metal fabrications that are subsequently welded together at extra cost by an unmachined aluminum die casting. Often, complication is added to reduce assembly costs, but these changes are made by the engineers involved in the task force, not by the computer program.

However, some aspects of DFMA will not suit all forms of manufacture. For example, IBM, which is a major user of DFMA, does not accept all the tenets of the system. As a general rule, Boothroyd and Dewhurst — and the developers of other approaches to DFMA — state that components should be assembled vertically, and preferably stacked on top of one another — uniaxis assembly. Although IBM appreciates the benefits for manual assembly, it does not consider uniaxis assembly to be the best solution with automation, because it limits a special-purpose assembly machine to operating on one face of the assembly only, when it could assemble on two faces at once. Also, it considers that for low-volume applications, several simple parts are better than one complex one. Obviously, the optimum design depends on production volume and the manufacturing equipment available, but that in no way invalidates the approach — it merely reinforces the fact that the designer remains in control.

To show what gains can be made, Boothroyd and Dewhurst cite the case of a spindle in a stainless steel bracket. The bracket consists of a folded plate with two parallel triangular walls folded up from the base with two flanges folded down for the fixing. There are a number of screws and bushes as well. This is an assembly found in many engineered products.

Boothroyd and Dewhurst analyzed the product and decided that since the bushings were nylon, it was preferable to produce an injection-molded nylon housing with integral bushes. With careful design of the spindle and form around the bushings, the need for fasteners could be eliminated. Although the original design looks simple, it consists of 10 components and several assembly operations, so it was rated with an efficiency of 7 percent. The redesigned assembly had no fasteners and only one simple assembly operation, so it had a very high rating — 93 percent.

Mouse Redesigned

A good example of DFMA is provided by the computer pointing device, commonly called a mouse, developed by the Video, Image & Printer Systems Group of Digital Equipment Corporation at Westford, Massachusetts. A team was assembled to handle the project, and the members started by assessing competitive products. They found that the main problem in assembly was the ball and cage in the housing. As the user moves the mouse, the ball rotates, acting as a sensor and transferring motion to the electronics. The team found that not only was assembly of these units complex, but that considerable adjustment — the bane of production engineers — was required.

Normally in a project of this type, efforts would have been made to simplify the cage assembly or to eliminate the need for adjustment, but because a CE task force was involved, a different tack was taken. Team members looked for an alternative sensor to replace the ball and cage. After an extensive search they found that there was such a device in the form of an inclined transducer, invented by Jack Hawley of the University of California at Berkeley.

Since the ball, cage, and ball retainer were eliminated, only 15 seconds is needed to assemble the sensor unit, against 130 seconds for the ball and cage used by competitors. Other changes included the elimination of seven screws with the use of snap-fits. Therefore, the parts count was cut by 50 percent, and the number of assembly operations was reduced by 33 percent. Total assembly time was 4 minutes 37 seconds, against 9 minutes 52 seconds, a savings of 53 percent.

But it did not end there: with the aid of data on injection molding and materials, the team was able to reduce the volume of material required by 47 percent. The package for the mouse was also designed afresh, cutting the cost from 59 cents to 24 cents. The complete project, including tooling, took 18 weeks, about half the average at Digital, clear evidence that the combination of DFMA and CE pays big dividends.

Parts Count Reduced

Also with the aid of DFMA, Digital Equipment reduced the parts count of the VT1000 graphics terminal by 43 percent, and the cost of components by 72 percent. The design cycle was reduced by 73 percent. The VAX 6000 minicomputer also benefited from CE and DFMA: the parts count fell by 20 percent, and assembly time was reduced by 42 percent. In both cases, the replacement of fasteners with snap fits played an important part in producing the gains.

NCR is also a dedicated user of DFMA, which it used to make dramatic changes in the way it builds its cash registers. It concentrated on eliminating fasteners and complex assembly jobs,

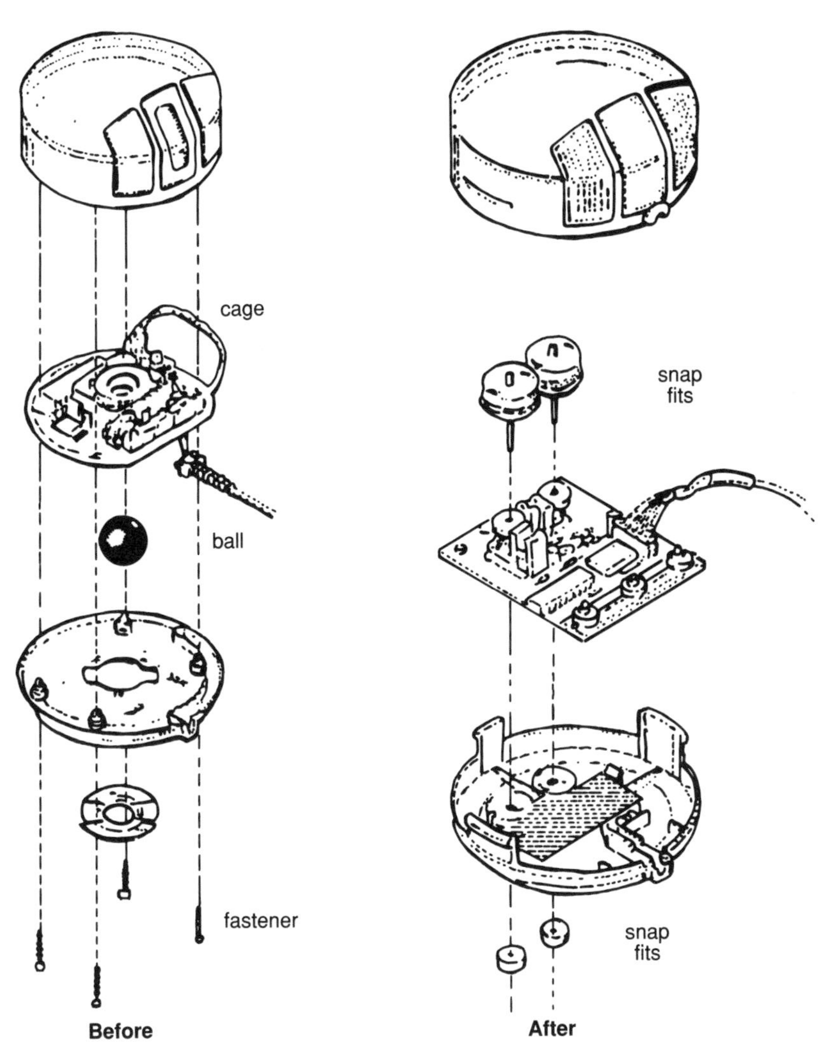

Source: Reprinted with permission from *Industry Week*, April 16, 1990. Copyright Penton Publishing, Inc., Cleveland, Ohio.

Figure 9-2. Digital Equipment's new mouse is notable for the use of fasteners and the absence of the ball and cage.

with the result that it has demonstrated an operator assembling a cash register blindfolded and without special tools. The results of the changes are spectacular:

- Parts count cut by 80 percent
- Assembly time cut by 75 percent
- Number of suppliers cut by 65 percent

The bottom line is that the savings over the whole life of the product are put at $1.1 million. Although that is substantial, Ford Motor Company, probably the first manufacturer to take up Boothroyd and Dewhurst techniques in a big way — it has trained 10,000 engineers in the methods — reckons it saved $1.1 *billion* in the first four years' production of the Taurus, the first automobile on which it used DFMA widely. Typical of the work by Ford was on a windshield wiper assembly, which was redesigned so that the number of parts was reduced by 36 percent, and assembly costs by 65 percent.

Suitable for Big or Small Volumes

Different industries have their own ways of approaching manufacture; what is suitable for an automobile would be prohibitively expensive for a small machine tool maker. The defense industry also has a set of peculiar requirements in manufacture: volumes are very low, while strength and durability must be high; and ease of service in the field is vital. The result is that many components are machined from the solid. The cost and waste involved are normally justified by the low volume.

Therefore, it is not surprising to find that the reticle assembly designed by Texas Instruments' Defense Systems and Electronics Group for use in an armored vehicle was extremely complicated. It consisted of three components machined from the solid, a gear drive actuator, a multitude of springs as tensioners and various fasteners, shims and retaining rings — a real nightmare for the assembly shop.

A producibility engineering group was formed by the company, and it applied DFMA to the reticle assembly and came up with a much simpler design. First, the end bracket was incorporated into the main housing, which was produced as an aluminum casting. That immediately eliminated two fasteners, some Loctite adhesive, and the bracket. It was realized that a cam could do the same job as the gear, and that it could be made with an integral knob. That eliminated two more components.

The reticle carriage was simplified in form and was made as a casting instead of being machined from the solid. It is now tensioned by two springs instead of nine, while shims and retaining rings have been eliminated. It has to be admitted that any engineer schooled in a mass-production industry would have been able to reduce some of the complexity of this assembly without the aid of DFMA, but that is not so easy in an environment where machining from the solid is considered appropriate for many housings.

In the event, the team at Texas Instruments made dramatic gains in costs while reducing the weight of the assembly by more than 45 percent. Of course, some tooling was required for the castings, but with such savings elsewhere, the overall benefits were still considerable.

Table 9-2. Improvements Made to Reticle by Texas Instruments with DFMA

Feature	Original Design	New Design	% Reduction
Assembly time, minutes	129	20	84
Number of different parts	24	8	67
Total number of parts	47	12	75
Number of operations	58	13	77
Fabrication time (hrs.)	12.63	3.65	70
Weight (lbs.)	0.48	0.26	45

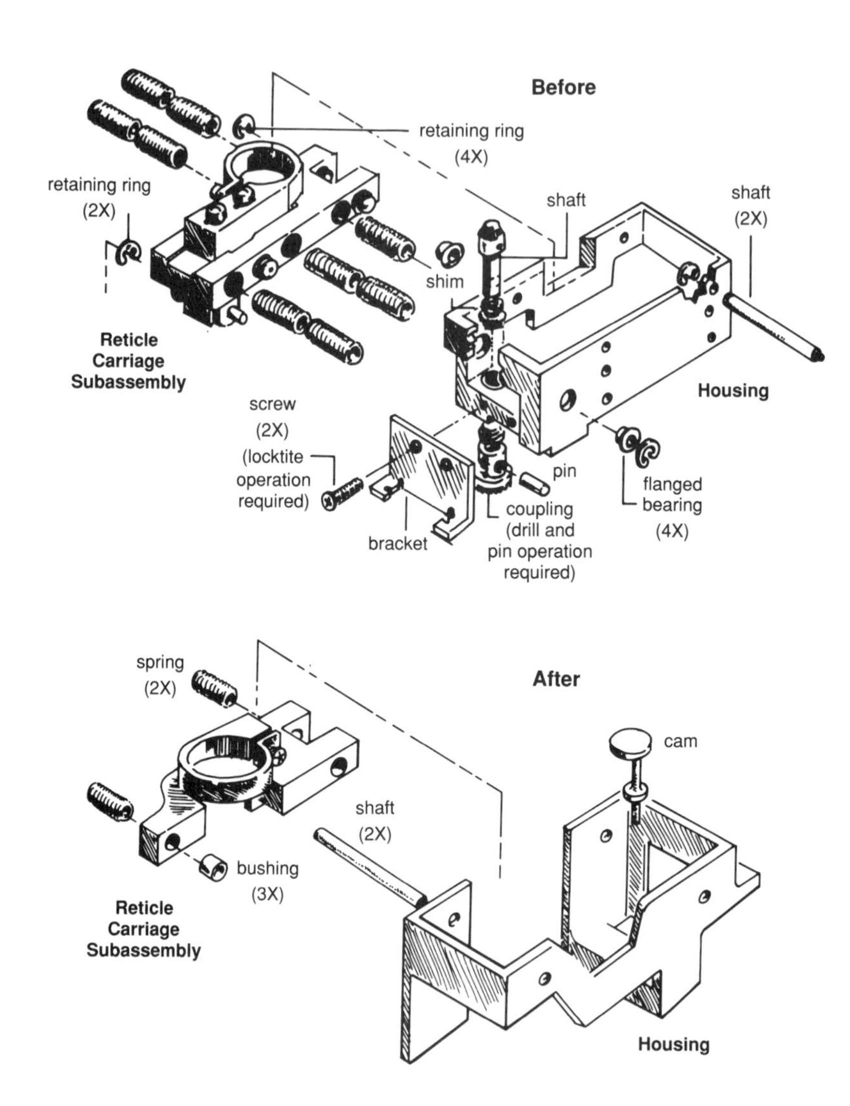

Source: Reprinted with permission from *Industry Week*, Sept. 4, 1989. Copyright Penton Publishing, Inc., Cleveland, Ohio.

Figure 9-3. Texas Instruments' Defense Systems and Electronics Group made big savings when it applied DFMA to a reticle assembly.

Fewer Assembly Operations

Another typical example was the experience of Ingersoll-Rand Portable Compressor Division (Mocksville, North Carolina). It adopted DFMA and CE and as a result cut lead time from 24 to 12 months. It hired Munro & Associates to train 34 people from design, manufacture, and marketing in Boothroyd and Dewhurst techniques, and split them into CE teams to evaluate various assemblies from compressors. The teams reduced the parts count from 80 to 29, while the number of assembly operations required fell from 159 to 40. These assemblies were actually put into production eight months later, and so impressed was the company that it put much more weight behind its CE program; it now has six CE teams operating on new product development.

In one of the follow-on projects, Ingersoll-Rand applied DFMA to a pair of separate panels — one for instruments and the other for controls. Each consisted of a recessed panel, one carrying two instruments and the other a relay, starter switch, and two-way valve. The control panel had a hinged door, whereas the instrument panel had a window. In the new design, the two panels are combined into one, and the hinged door remains. As a result, the number of fasteners was reduced by 38 percent, and total parts count is reduced from 36 to 24. Since there are 30 assembly operations instead of 45, assembly time is down 28 percent to 6.1 minutes.

Printed Circuit Boards

In addition to the general-purpose DFMA programs, Boothroyd and Dewhurst have developed a special program to estimate the cost of assembly of printed circuit boards (PCBs). It includes a data base for inserted costs of most components in use and a number of equations. The decision starts with the type of PCB to be used, since the overall cost will be influenced by the method of assembly — whether the bulk of components are inserted or surface mounted, for example. Then, the type of components is assessed, and the functionality of the PCB assembly is considered. The greatest assembly cost is incurred with the assembly of jumper wires and non-standard components that are attached by fasteners, rather than being soldered in place or inserted through holes in the PCB.

Not surprisingly, the use of automated insertion usually costs far less than manual operations. For example, a standard component can be inserted automatically in 2.4 seconds, whereas manual insertion takes around 6.0 seconds. Equivalent times for a radial IC are 10.0 and 4.0 seconds, respectively, according to Boothroyd and Dewhurst. Since insertion machines are extremely flexible, and can insert many different components of similar form, their use can be easily justified.

In the assembly of a logic board containing 69 dual in-line packages, (DIPs) 1 SIP socket, 16 axial components, and 32 radial components, the program estimated the assembly cost as $4.61. Had the 32 radial components been inserted manually, the cost would have increased to $6.80, up almost 50 percent. Equally, the elimination of 11 fasteners used to secure 2 components would have reduced the cost by $0.48, or about 10 percent. The ability to quantify these benefits on a comparative if not absolute basis improves the team's ability to reduce costs where most gains are to be made.

Although Boothroyd and Dewhurst are best known for their work on assembly, they have investigated the cost of machining parts from bar on CNC turret lathes. This showed, for example, that on relatively large parts, the material cost usually amount to 80 percent of the machined part cost. Therefore, changes to tooling to speed up cutting hardly affect the parts cost at all, especially since machining costs represent only 20 to 25 percent of the total part costs. However, with very small components of volumes of less than 0.5 in^3, the situation is very different; the manufacturing cost accounts for more than 75 percent of the machined cost. This study was used as part of the machining program Boothroyd developed.

Injection Molding

Also proving its worth is a Boothroyd and Dewhurst program to evaluate injection moldings called "Parts Cost Estimating — Injection Molding." It takes into account the material, the type of mold, the features, and quality. When the component is analyzed by the program, it determines first how many cavities should be machined into the mold. The cost of the cavity and cores, the production volume, and molding cycle time are all evaluated, and the size of machine is selected. A relative operating cost and dry cycle time are also calculated. Subsequently, the cost of materials and the time taken to fill the cavities during injection are calculated. The program incorporates modules to allow users to indicate the surface area and volume of the component simply.

In a typical example, a heater core cover was being molded for one of the automakers in a mold with six cavities. The six cavities were needed to match the production volume requirements with the cycle time. The program estimated the cost of the mold at $36,380, against an actual cost of $38,750, an indication of the accuracy provided.

The covers were produced on a machine with a running cost of $72 per hour, and the cycle time was rather long, at 42.8 seconds. The long cycle time was necessary because there were thick pads at the lugs for the fasteners and the thickness was at least twice that of the walls of the main molding. In addition, there was an abrupt change of section at the joint between the pads and the main housing.

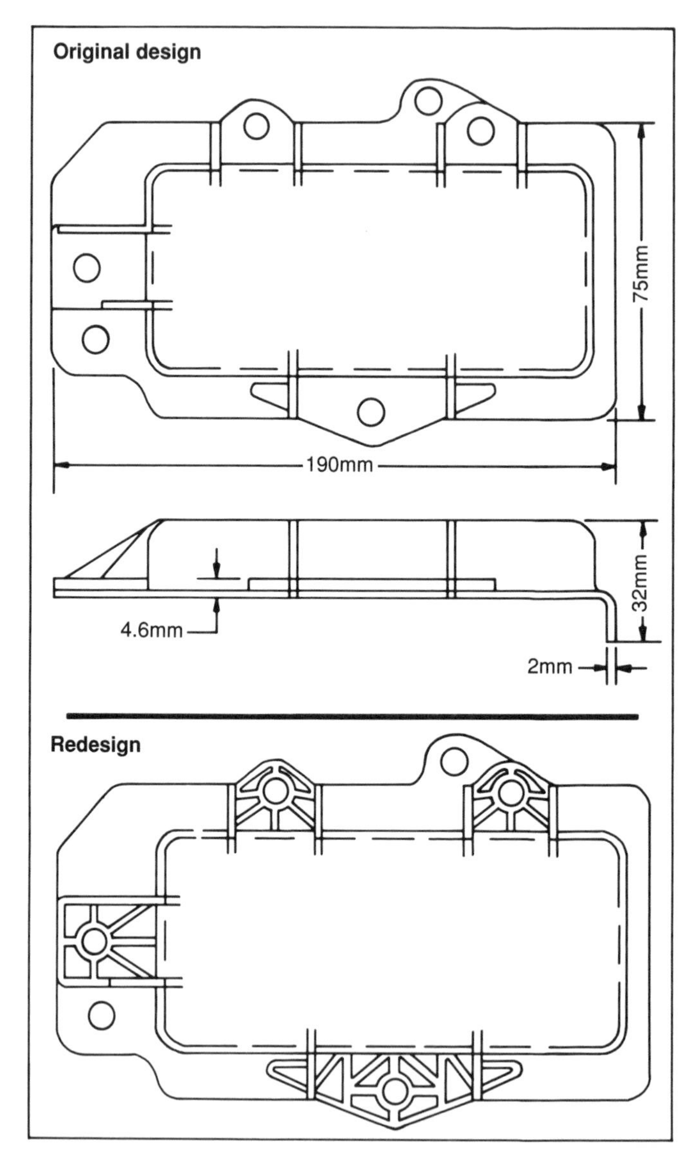

Source: Reprinted with permission from *Machine Design*, July 21, 1988. Copyright Penton Publishing, Inc., Cleveland, Ohio.

Figure 9-4. Boothroyd and Dewhurst changed the design of this cover by replacing the thick pads with ribbed bosses and reduced the molding time but maintained stiffness.

Although the simple form of thick pads projecting from flat walls does reduce mold costs, the thick sections result in sinkage and poor quality. Since this is bad practice for a molding or casting, the cover was redesigned with a number of ribs replacing the pads, so that the wall thickness remained constant throughout the molding. Another simulation was run by the program, with dramatic results — the cycle time would fall from 42.8 to 13.3 seconds. Therefore, a two-cavity mold could be used instead of the one with six cavities, reducing overall mold costs, despite the more complex machining. The program showed that the parts cost could theoretically be reduced by 33 percent.

Table 9-3. Estimated Cost Reduction with Redesigned Heater Core Cover

	Original Design	Modified Design
Cost of production mold	$36,383	$22,925
Cycle time, seconds	42.8	13.3
Number of cavities	6	2
Cost of component	25.1 cents	16.8 cents

However simple or complex the molding or assembly, DFMA will help cut the costs of assembly and reduce parts count, which reduces costs in all departments — from purchasing to replacement parts. However, it is only completely successful when used by a task force in a concurrent engineering regime; otherwise, the changes will almost certainly be made too late.

Failure Modes and Effects Analysis (FMEA)

Another useful technique in eliminating poor design features is failure modes and effects analysis (FMEA), or failure modes and effects criticality analysis (FMECA), as some exponents prefer to call it. Chrysler is among those corporations that built FMEA into its CE system, while Delco Remy uses a similar technique.

The aim is to highlight the areas or assemblies most likely to cause failure in the complete assembly. For example, when an engine fails, it is extremely rare that the cause is a breakage in a component such as a connecting rod or crankshaft. The most likely cause is a fault in the ignition or fuel system. With most electronic assemblies, faults occur most often in connections and ancillaries.

In both cases, FMEA helps assess the importance of the failure in these small components. For example, in the engine, it might be found that the replacement of a tiny component costing 20 cents with an apparently overengineered one costing 35 cents eliminates more than 40 percent of recorded failures. With FMEA, it immediately becomes apparent why the cost of a tiny component should be almost doubled; without it, management is unlikely to accept the change.

FMEA defines function as the task a component performs — the function of a vale is to open and close, for example — and failure modes are the ways in which it can fail. The valve will fail to close if its spring breaks, but it can also stick in its guide or be held open by the cam should the camshaft belt break.

The technique consists in assessing three aspects of the system and how it operates:

- Anticipated conditions of operation, and most probable failure
- Effect on performance of the failure
- Severity of the failure on the mechanism

The probability of failures is usually rated on a scale of 1 to 10 with the higher number being more critical. The difficulty of detecting a failure before the assembly is in use is also included in the equation, with a value of 1 indicating easy to detect, and 10, very difficult. A criticality index or risk priority number can be produced:

$$C = P \times S \times D$$

where C is criticality index; P is probability of failure; S is the seriousness of the failure; and D is the difficulty of detecting failure before assembly.

This technique is useful for rating alternative solutions to a problem but is not easy to use with accuracy for new designs. Therefore, GE devised a system especially for components for which no reliability data are available.

Quasi-Tree to Spot Problems

Whereas FMEA usually requires either a worksheet or tree approach, GE produces what it calls a quasi-tree.* The starting point is the event that is to be avoided — the top event — which, in the case of an engine, would be that it stops, and in most assemblies is a failure to operate for whatever cause. Next in the hierarchy below the top event are the major assemblies, and then the components that make up these assemblies. Beneath the components are the failure modes that emanate from them. As with the normal FMEA technique, each component and each failure mode is rated. GE found that for this procedure, a multi-disciplinary team produced the most accurate evaluation — another reason for using CE.

GE ranks the assemblies and parts on a scale from 1 to 3 according to their likely impact on failure — the higher number indicating a minor effect and the low number a major effect. The

* R. E. Warr, *Failure Modes and Effects Analysis Method for New Product Introductions,"* SAE Technical Paper 841600, 1984.

assemblies and components are then weighted, with the sum of the components in one assembly being equal to the total for the assembly. The total for all the assemblies is 100. Therefore, in a case where there are three assemblies, weighted at 20, 30, and 50, the three components in the first assembly must add up to 20, and so on. Similarly, with the contributions to failures, the sum of the failure modes for each component must equal 100 percent.

GE adopted a software program called PREDICTOR to gain results from the data speedily. In the case of a slide projector, the three relevant assemblies were the fan, lamp, and switch. Typical failure modes were breakage of the belt, an open filament in the lamp, and the switch being stuck open. The analysis revealed that the root causes of failure, and their criticality were

- short to ground, 29 percent
- open filament, 19.8 percent
- switch stuck open, 16.9 percent
- switch stuck closed, 10.3 percent

FMEA is useful in assessing whether there are an unnecessary number of components in an assembly since the interaction of one assembly on another will multiply the failure effects. It is equally helpful in analyzing the product and the equipment used to produce the assembly.

FMEA is complementary to both QFD and DFMA, and assists in the identification of the failure modes that are likely to cause product failure in use. It also helps eliminate weaknesses or overcomplication of the design, and identify the components that are most likely to fail. Its use should not be confined to the product under development by the CE task force. It can also be used effectively to assess causes of downtime in the machines in the production shop — before their design is completed.

Table 9-4. Relative Rankings, Weights, and Percent Failure Contributions for Assemblies and Components

Item	Relative Rank Assembly Part	Relative Weight	Percent Failure Contributions
Assembly 1	3	60	
Part 1	2	35	
Failure mode	1		60
Failure mode	2		40
Part 2	1	25	
Failure mode	3		35
Failure mode	4		45
Failure mode	5		20
Assembly 2	2	40	
Part 3	1	20	
Failure mode	6		63
Failure mode	7		37
Part 4	1	20	
Failure mode	8		50
Failure mode	9		50

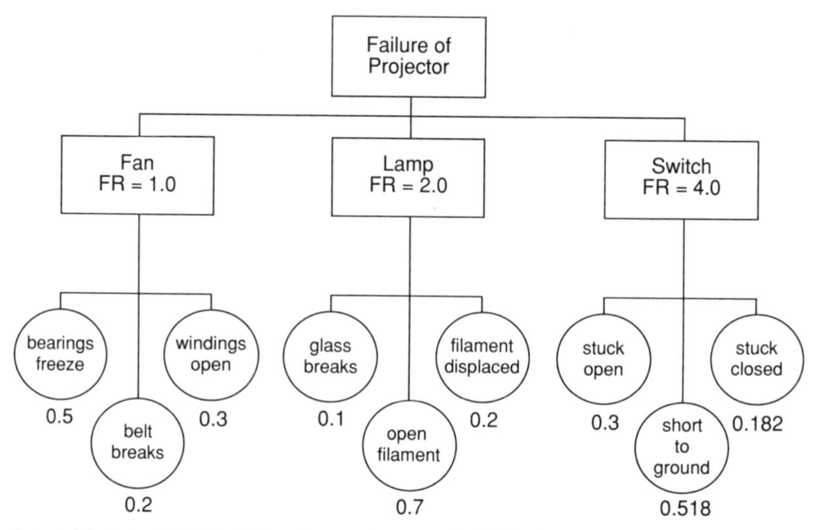

Source: R.E. Warr, General Electric Co., SAE Technical Paper 841600. Reprinted with permission ©1984 Society of Automotive Engineers, Inc.

Figure 9-5. GE's quasi-tree shows the results obtained with FMEA on a slide projector.

CHAPTER TEN

Robust Products for Manufacture

Taguchi system optimizes process efficiently
Success in wide range of industries and processes
Troubleshooting with fewer tests than normal

To refine the product further, you need to investigate problems likely to be encountered in production before they cause trouble. One valuable technique that can be used to optimize the basic parameters and check the effect of variations in manufacturing settings is that devised by Dr. Genichi Taguchi.

Originally developed to improve the productivity of design, Taguchi's approach can be applied equally well to optimize processes; indeed, most users in the West have applied it to processes rather than to design. Dr. Taguchi developed the technique while at the Electrical Communications Laboratory of the NTT, Japan's telecommunications company, which was state-owned until recently. Over the past decade, NTT has acquired an enviable reputation for the scope of its research and the speed with which it develops new devices and products.

Taguchi Optimization

When he started at NTT in 1950, Taguchi sought to simplify engineering design at the stage where a well-defined concept is turned into a detailed design. The aim was to provide a system to choose between the number of alternative specifications available. For example, there might be three different materials, two different material treatments, and three different structural shapes to be considered. To test prototypes of all variables would take a long time and be costly. Moreover, the production process might involve a number of variables; in the case of a die casting, factors such as material specification, shot weight and temperature, die temperature, and pressure should be considered.

Without Taguchi, problems in the early stages of manufacture can lead to a high scrap rate and the substitution of a different material and method, whereas subtle changes in the process might solve the problem at little cost or delay. Taguchi's basic approach is that the design must be made robust — that is, able to be produced to good quality despite the variables inevitable in the manufacturing process.

Taguchi's work has been taken up by many corporations. The American Supplier Institute (Dearborn, Michigan) has gone so far as to develop training techniques in Taguchi methodology and register "Taguchi Methods" as trade and service marks.

Off-line Quality Control

Taguchi's approach is usually called "off-line quality control" because it takes the biggest responsibility for quality control away from the production line. To Taguchi, quality involves design, manufacturing processes, production, and the performance of the product in service. He defined the quality of a product as "the loss imparted by the product to the society from the time the product is shipped." In other words, any requirement for maintenance, or any fault that either inconveniences the user or requires rectification reduces the quality of the product. The ideal product would be one that never required any attention, continued to perform adequately when worn, and was recycled when completely worn out.

Taguchi set out to improve the productivity and quality of design to turn quality into an inherent factor of a product rather than a discipline added at the manufacturing stage. His first principle is parameter design, in which the basic parameters of the design are established. For example, in injection molding, shot weight and temperature are among the parameters. Next, ways of achieving optimum levels of these parameters are considered. The object is to achieve a design that will be produced naturally at or near the nominal design requirements.

Reducing Variations in Manufacture

Equally important are the variables in manufacture, such as moisture and dust, that can lead to defects being produced. These are considered separately from the parameters. Taguchi calls these "noise," and adopted the concept of signal/noise ratio to indicate whether the result is acceptable or not. For example, in the case of strength or yield, the signal/noise ratio should be high; for contamination it should be very low.

Taguchi also evolved the concept of the loss function. He pointed out that imperfections in a product result in a loss to society on account of the need for service or replacement parts. He argued that the loss could be quantified in terms of the loss to the manufacturer — warranty claims, and so on — and to the customer in lost opportunity or service costs. The loss function allows the cost of eliminating the fault to be quantified.

This particular concept has not received widespread use. For example, Honda considers that "a fault is a fault, is a fault" and therefore must be eliminated. However, some firms have adopted Taguchi's formula to calculate the value of eliminating a fault in a product to the customer.

Taguchi Methodology

Taguchi's work does not involve vague concepts; he defined methods of proving designs, and these are attracting much attention and yielding good results. For example, between 1983 and 1987 ITT trained 1,200 engineers in Taguchi methods, and studied 2,000 cases with the aid of Taguchi methodology. The result was a savings of $35 million a year, and in only 10 percent of the cases was some capital required to help produce the savings. AT&T Bell Laboratories also has considerable experience in the use of Taguchi methodology.

Orthogonal Arrays

In practice, Taguchi methodology starts with his orthogonal arrays. These developed the idea of analysis of variance (ANOVA), evolved by R. A. Fisher in the United Kingdom in the 1920s. The aim was to develop a method of testing a number of variables in the minimum number of tests. For example, Taguchi's *L9* orthogonal array requires only 9 tests against 27 for Fisher's Latin square technique, and 81 for a full test of all possible permutations, known as the full factorial, in this case 3 × 3 × 3 × 3. This reduction is of considerable benefit in the real world of engineering and can make the difference between a series of tests being conducted, and being avoided as too costly — with the result that a simple solution is missed, and either products continue to be scrapped and reworked, or a new

Table 10-1. Typical Taguchi Orthogonal Arrays

Array	Number of factors and levels	Number of trials in full factorial
L4	3 × 2 levels	8
L8	7 × 2 levels	128
L9	4 × 3 levels	81
L12	11 × 2 levels	2,048
L25	6 × 5 levels	15,625

approach is adopted. The real cost of either is far greater than that of adopting Taguchi.

The number after the *L* in the Taguchi orthogonal array indicates the number of tests required; an *L4* requires only four tests. As an example, consider a case where there are four parameters, with three levels for each. The parameters might be injection speed, temperature, and pressure, and different formulations of the material. The levels might be low, medium, and high. If a full factorial approach is adopted, 81 tests would be required: $3 \times 3 \times 3 \times 3 = 81$. If the Taguchi *L9* array is used, only 9 tests are needed.*

Parameter Design

Tests of various combinations are conducted, and then the transients — Taguchi's noise — that occur in manufacture, such as humidity, lubricant, and dust, are tested as well. Although every interaction may not be evaluated, those that could be critical are usually shown up by the tests. This occurs because the tests are balanced, providing suitable relationships between variables. However, extra tests may be required in some cases, but if so, the original set of results will indicate which tests are needed. Once the best settings are found a batch should be run to replicate the results.

When the complete design is considered, it is necessary to take into account external noise — transients involved in the process — and internal noise, which includes deterioration during storage and wear through use. Then, there is variational noise, which is the difference between individual products made to the same specification.

---

* For more detail on Taguchi and on statistical analysis, see A. Bendell, J. Disney, and W. A. Pridmore, eds. *Taguchi Methods: Applications in World Industry* (Kempston, England: IFS Publication/Springer Verlag, 1989) and G. Wilson, R.M.G. Millar, and A. Bendell, *Taguchi Methodology with Total Quality* (Kempston, England: IFS Publications, 1990).

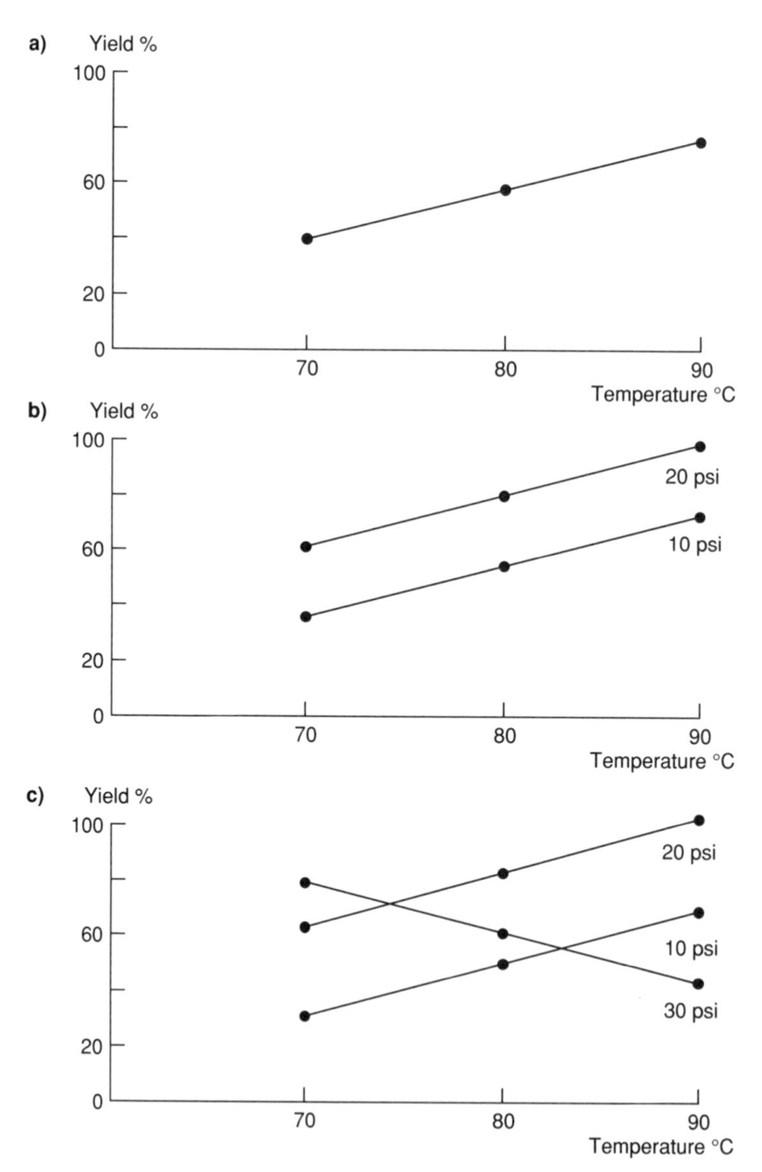

Source: G. Wilson, R.M.G. Millar, and A. Bendell, *Taguchi Methodology with Total Quality* (Kempston, England: IFS Publications), p. 14.

Figure 10-1. Tests with different levels of the relevant parameters indicate optimum settings.

Brainstorming

Before the tests are conducted, it is of course essential to establish the parameters. The task force will do this at a brainstorming session — brainstorming is considered essential in the exploitation of Taguchi's approach, as in most aspects of CE. However, this is disciplined brainstorming, with everyone bringing as much information on the product and process as is known. Therefore, each company will need to exploit its experience in design and manufacture to the full. It is also one more reason why the task force approach is important.

Once the parameters and noise have been established, the tests are determined. It is essential that all information is supplied to people in written form and that procedures are adhered to exactly. In many cases, it has been found that the settings for the parameters chosen as the optimum by the Taguchi tests are quite different from any used before.

Analysis of the test results can take one of two forms; statisticians in the West favor the use of ANOVA statistical techniques, but Taguchi himself advocated visual analysis of graphs, which show relevant interactions clearly. He argued that it is easy to see from the curves where variations were significant, and this is an approach that will appeal to engineers. It has the merit that all members of the task force can analyze the results; when an analysis is presented by the statistician, some members may feel that the results are unintelligible.

Gains in Many Industries

Taguchi's approach can be used successfully to identify the optimum design or to prove or eliminate problems from the process. It has proved invaluable in optimizing processes in semiconductor manufacture, where there are many different processes involving different disciplines in sequence and where the target is always to reduce the size and therefore variation in components. It has also rescued many a project involving plastic injection molding.

Ceramic Tile

One interesting example was in the manufacture of ceramic tiles. A Japanese company had purchased a large baking kiln at a cost of some $500,000, and found that 30 percent of the tiles were oversize. After some tests and discussions, it was decided that the problem was the temperature gradient between the center and sides of the stack of tiles. Possible solutions were to buy a new kiln, or to modify the kiln to reduce the temperature gradient, or to modify the design of the tile, using Taguchi's approach.

Following a brainstorming session, seven factors affecting size were identified, and a series of tests run. It was found that increasing the amount of limestone from 1 percent to 5 percent reduced the proportion of rejects to less than 1 percent. It was also found that the amount of agalmatile, the most expensive ingredient, could be reduced, so the result was improved quality at lower cost, without any change in the equipment.

Semiconductor Processes

In the U.S. semiconductor industry, Taguchi's methodology has been adopted to optimize many aspects of the process, including growth of epitaxial layers, plasma etching, and the forming of small windows in integrated circuits (ICs). The work on the windows in CMOS ICs undertaken by AT&T Bell Laboratories is interesting because a test of all permutations would have required 5,832 tests; the Taguchi orthogonal array required just 18 tests.

There are five critical processes involved in the forming of the large number of 3.5 micron diameter holes, 2.0 micron deep in ICs. The tests revealed that

- the mask used to expose the photoresist for the hole should be increased from 2.0 to 2.5 microns
- the speed of spinning the wafers to produce a uniform thickness of photoresist should be increased from 3,000 to 4,000 revolutions per minute
- the developing time in which the photoresist is dissolved should be increased from 45 to 60 seconds

The result was a faster process and better quality, whereas conventional wisdom often suggests that more time is needed to improve quality. Also, in this case changes were needed at several stages of the process, so that without Taguchi, a very complex series of tests would have been needed.

In another case, AT&T Bell Laboratories used Taguchi to grow epitaxial layers on silicon wafers. The wafers are mounted on cylindrical susceptors, which have seven sides and are rotated in metal jars. Gases enter the jars and are deposited on the wafers. The problem with this approach was that the layer was not within the 14 to 15 micron thickness required on all wafers.

After a brainstorming session, it was decided that the main parameters were

- nozzle position
- rotation, or oscillation, of susceptor
- temperature
- time
- arsenic flow rate
- hydrochloric acid etching temperature
- hydrochloric acid flow rate

Two settings were required for each parameter, and 16 tests were run. The mean and variance of epitaxial thickness were calculated for each run, and it became clear that continuous rotation produced better results. The tests also showed that the nozzle position was critical. Once continuous rotation was adopted with a new position for the nozzle, actual variations in the thickness layer were reduced by 60 percent, and the time of deposition adjusted to produce a mean level of 14.5 microns.

Other examples in the semiconductor industry include ITT Cannon, which used the technique to improve the uniformity of the layer of gold plating on electrical contacts, and Intel, which used it to optimize plasma etching.

Optimized Computer Responses

AT&T Bell Laboratories also used the technique to optimize the response time of a Digital Equipment VAX 11-780 computer running on the UNIX operating system, demonstrating that the method is appropriate for use not just in computers, but to optimize the operating rate of many different mechanical systems.

Eight parameters were involved, with three levels of each. With a full factorial permutation, 2,916 tests would have been needed. With the Taguchi technique, 18 tests were made, despite the fact that the use of different hardware and software were considered. The eight parameters were

- disk drive configuration
- distribution of key system and user files among disks
- memory size
- system buffer size
- number and selection of "sticky bits," a system that speeds up execution by keeping all commands together in a single area of the disk for rapid recall
- number and size of computer-to-computer terminal communication devices — KMCs

Each test was run by checking the computer response time regularly for 2 1/2 days, the tests being spread over 12 weeks. The tests were illuminating. For example, with 3-megabyte

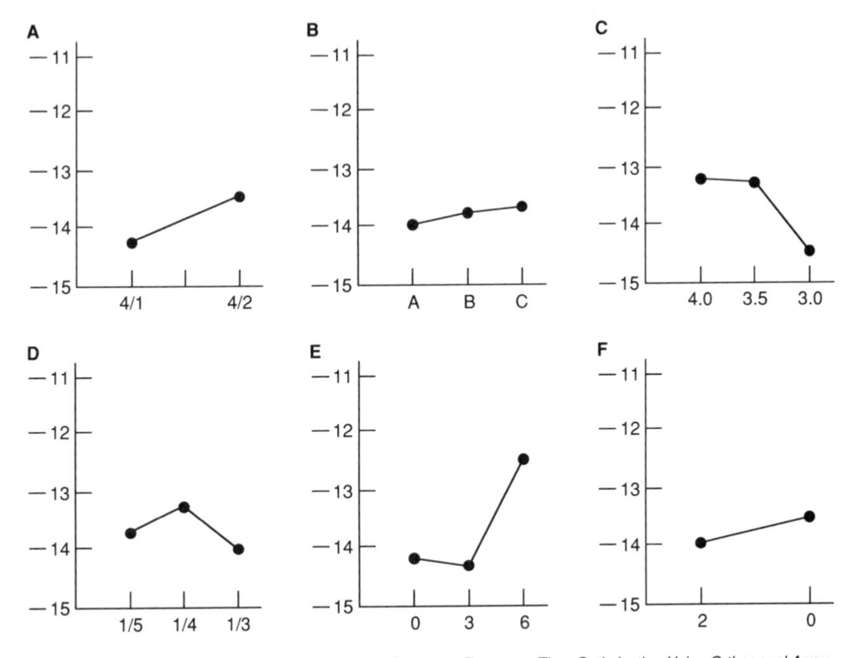

Source: T.W. Pao, M.S. Phadke, and C.S. Sherrerd, *Computer Response Time Optimization Using Orthogonal Array Experiments*, 1985 IEEE International Conference on Communications, June 23-26, pp. 890-895. ©1985 IEEE.

Figure 10-2. Analysis of the main parameters in optimization of a minicomputer.

memory, response time was improved, but an increase beyond that level would make little difference. An extra disk drive would also improve response, but the KMCs, designed to improve response, did not help at all in this application. Overall, with the optimum configuration, response time was reduced by 60 percent. The changes were of a minor nature:

- A combination of one disk drive with four disks and another with two drives replaced one drive with four disks and one drive with one.
- A small change was made in the system buffering.

- Two communication devices intended to speed up response were eliminated.
- Eight "sticky bits" were incorporated.

Automobile Industry Benefits

Many companies in the automotive industry have gained from the use of Taguchi as well. Flex Products, a vendor to GM, was experiencing difficulties in producing a speedometer outer cable to close tolerances, because of shrinkage. Too many were being rejected — and only after the car was in service. The cable consists of a polypropylene liner, wire braid, and an extruded coating, and the main problem was excessive shrinkage of the coating. The result of the Taguchi tests was much less variation and lower shrinkage overall — a mean of 0.05 percent instead of 0.25 percent.

Using Taguchi's loss function, Flex Products was able to calculate the value of the change. Each complaint in the field, shown to occur with a shrinkage of 1.5 percent, resulted in a service cost of $80. These figures are used to calculate a value, K.

$$K = (80/1.52) = 35.56$$

The loss factor, L, is approximated as $K\theta 2$, where $\theta 2$ is the mean signal/noise ratio. Therefore,

$$L1 = 35.56 \times 0.0595 = 2.12, \text{ and } L2 = 35.56 \times 0.0037 = 0.13$$

The improvement equals a saving in loss function of $2.12 − $0.13 = $1.99. That is not the actual cost saved by the manufacturer, but is an attempt to quantify the loss saved on each product.

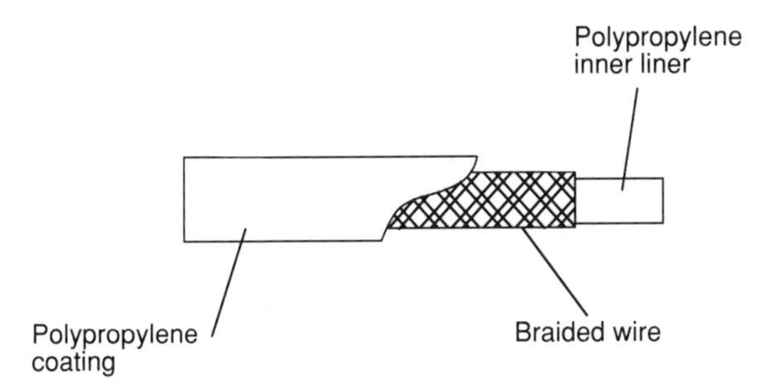

The product: extruded thermoplastic speedometer casing.

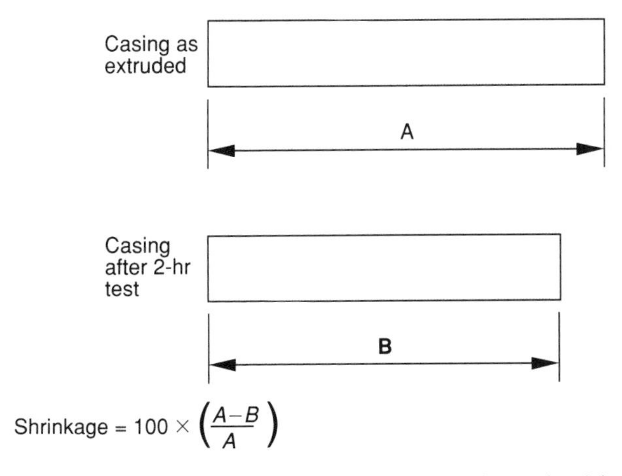

$$\text{Shrinkage} = 100 \times \left(\frac{A-B}{A}\right)$$

The quality characteristic: percentage shrinkage after 2-hour soak test.

Source: Flex Technologies, Inc.

Figure 10-3. The speedometer cable produced by Flex Products suffered from excessive shrinkage.

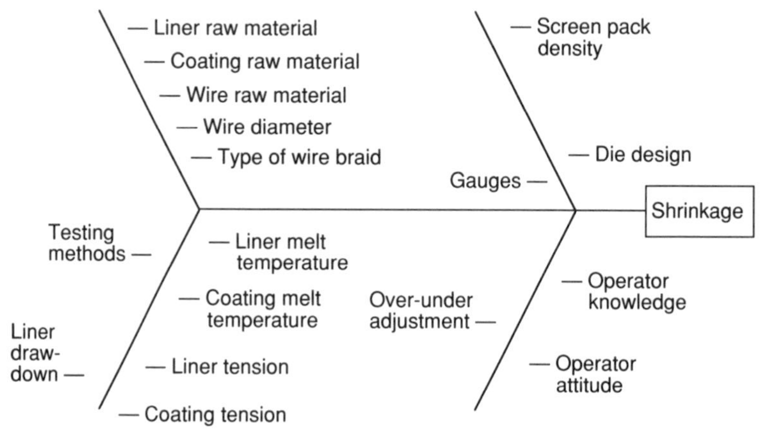

Figure 10-4. A cause-and-effect diagram was used to analyze the problem of shrinkage.

Voids in Moldings

Davidson Interior Trim Textron (Dover, New Hampshire), also a vendor to GM, had a major problem in the molding of urethane foam consoles installed in the rear of the Buick Riviera. The problems arose from the thin section of the molding, which resulted in voids, soft spots, and underfills. These faults resulted in a reject level after molding of 47 percent; even after costly reworking, 13.5 percent of moldings were being scrapped for these reasons.

In this case, the parameters were

- shot weight
- mold temperature
- the condition of the insert
- venting of the mold
- the viscosity of the spray wax used

A total of 32 tests was run, and the results showed that better results would be obtained if the mold temperature was reduced, and the vent eliminated. Immediately, a confirmation trial was run — this is the standard procedure with Taguchi tests — and despite operation during different shifts and with different shells, the results confirmed the experiments.

The final results for all shifts showed that rejects were now caused principally by a folding or rolling of the edge, at 23 percent, whereas voids were found on only 14.5 percent of moldings. In fact, 16.2 percent of moldings were rejected at this stage, and after reworking, the scrap level caused by voids, soft spots, and underfilling was reduced to 2.6 percent.

Since venting is not required, the labor cost is reduced, and the lower temperature reduces running costs. Overall, therefore, Davidson Interior Trim Textron calculated the value of the savings as $14,225 annually.

Also suffering from poor moldings, but in this case solid housings for its copiers, was Rank Xerox. There were unacceptable variations in dimensions, surface finish, and curvature on an injection-molded cover, and the total number of possible permutations was huge. However, the Taguchi methodology allowed the use of the *L*18 array, giving only 18 tests. The approach was highly successful, with the tests identifying the parameters that would allow these conflicting requirements to be met. Significantly, the machine settings were quite different from those that had been run previously, a common result of Taguchi experiments.

One vendor that was about to lose a contract to supply chromium-plated trim components produced from 434 stainless steel also found a rescuer in Taguchi. This steel is difficult to plate, and Production Anodizing Company could not obtain the desired thickness nor the bluish color required by Chrysler

Corporation. The components were actually assembled into the products supplied to Chrysler by ITT Automotive, so all three companies attempted to solve the problem together.

The aim was to produce a plating 0.003 to 0.01 inches thick. There were some difficulties in deciding on the parameters — in fact, they decided that time should elapse before they finalized the list, to give time for each member to firm up his thoughts. Eventually, an *L*12 array was adopted, and after the tests were run, there was some conflict between optimum conditions, verifying that the process was difficult. Despite these conflicts, a set of conditions was selected that produced satisfactory plating, with all components within the permitted range, whereas previously all had been outside it. The average proportion of acceptable products was 86 percent.

Had the plater not been able to resolve the problem, it would have lost business worth $298,000 a year; had ITT Automotive lost its contract, it would have lost business worth $4 million a year. So Taguchi really did rescue these companies.

Quite different was the problem encountered at the Cadillac assembly plant, both in terms of its type and what the Taguchi tests showed. There was a niggling problem in which every so often, an automobile would be found to have its headlights out of aim, not long after the aim had been adjusted. It was determined that the slamming of the heavy hood at the end of the line was causing the maladjustment.

The headlight was retained by a pair of preset screws and a spring that pulls the lamp against a Delrin pin. Initially, there were thought to be 24 parameters, but the list was narrowed down to

- whether the screws should be preset or not
- whether to use the Delrin pin or not

- whether to lubricate the pin or not
- what strength spring to use

An *L*16 array was adopted, and a total of 64 observations was made. The results confirmed that the presetting of the screw, the Delrin pin, and strength of spring were all significant. Indeed, it was found that the Delrin pin could be eliminated without detriment to the retention of the setting, and that best results were obtained with a light spring, some lubricant, and a preset screw. Whereas previously 10 percent of headlights had been at least two inches off aim, and almost all were biased to one side of the target, the modifications resulted in a much lower variation, with virtually all within one inch of the target, and 20 percent were virtually on the dot.

Summary

All these examples demonstrate that Taguchi's approach is a very useful tool in optimizing the design, the process, or the variations in assembly. Only a minimum of testing is required. Taguchi's system can be taught so that a standard approach is adopted throughout an organization, however large or small. It fits in neatly with the task force system of project management and will almost certainly bring about lower costs and more thorough testing than other methods.

CHAPTER ELEVEN

Concurrent Engineering Is Wasted without CAD/CAM

Coarse simulation at concept stage
Tools and dies produced from CAD data
Precursor to genuinely integrated manufacture

To maximize the benefits of concurrent engineering, the current trend toward increased computer-aided design and manufacturing (CAD/CAM) must be exploited. With the right combination of hardware and software, design and stress engineers can work in parallel, far fewer prototypes need to be built, and lead times can be cut dramatically.

Even without an integrated system, the use of CAD allows manufacturing engineers and the suppliers of machines and components to see the real product, whether it is still at the concept stage or is a finalized design. Without CAD, there is too much margin for error; virtually every two-dimensional paper drawing leaves the person studying it some areas for interpretation, especially where compound curves are involved. With a three-dimensional computerized image, the dimensions of the product are complete.

Clearly, manufacturers will want to limit the number of outsiders having access to concepts of products not planned to reach the market for two or three years. For this reason, some companies are reluctant to transfer design data of their products to vendors. But in practice, this is little different from sending a drawing to a supplier, or a few cylinder blocks to the manufacturer of the machine tools for trials. In any case, the whole point of involving outsiders as members of the task force is that they will act as part of the team and will be subject to the same rules on confidentiality as company employees. Thus, with the combination of CE and CAD the manufacturer has a greater chance of maintaining secrecy for future products than with over-the-fence engineering.

Ultimate Aim

Developments in data bases and distributed systems are coming so fast that a long-term program for the storage and use of data can be drawn up. Using CE, the aim with data for new products is that

- relevant members of product engineering and manufacturing engineering have access to workstations in a network
- a common data base, maintained by product engineering, be accessible by all departments

In most cases, people accessing the drawing see the design data only, but they will soon be able to access the data in a form suitable for direct use in their department, whether that be for manufacture, service, or marketing.

Most CAD packages maintain a common set of data that can be used for design or manufacture, but there are limitations to its use in manufacture at present. However, common surface data can be available for manufacture in the following ways:

- Nesting of blanks for pressing and folding
- Direct machining of tools and dies for pressings, castings, and moldings

- Direct dimensions for fixtures for body welding and assembly
- Direct dimensions for molds and cores or castings
- Metal cutting of any component

Nesting programs are already in wide use, as are programs for metal cutting on NC machines. Some manufacturers, especially in Japan, are able to produce dies directly from design data, and this approach will be more common by the mid-1990s. Rapid developments are now likely because of the maturity of the main CAD software packages and the rapid advance in the ratio of computing power to cost. For example, the performance-to-cost ratio of a workstation designed to run three-dimensional design software is today nine times what it was in 1988. By 1992, Control Data forecasts that the power-to-price ratio will be 200 times that in 1988.*

Currently, a workstation for CAD use has a processing power of 12 to 25 MIPS (million instructions per second), which is up to 50 times that of a decade ago. With the use of RISC processors the power of the workstations is increasing, and in the spring of 1991 Hewlett-Packard introduced the HP Apollo Series 700 workstations with processing power of 57 to 76 MIPS, well ahead of the predicted trend line.

Although the basic processing speed is important, the main bottleneck is in the time taken to deliver the data to the screen, and in manipulating it onscreen quickly enough. The HP Apollo 700s operate at 1.1 million 3-D vectors per second — five times faster than good entry level Reduced Instruction Set Computers (RISC) machines — which also pushes performance upward to levels that others will soon emulate.

* I. A. Wulf and R. Sterbl, "Simultaneous Engineering for the Automotive Industry Using Workstations and Computer Servers," in *Proceedings of Auto Tech*, Birmingham, England, November 1989.

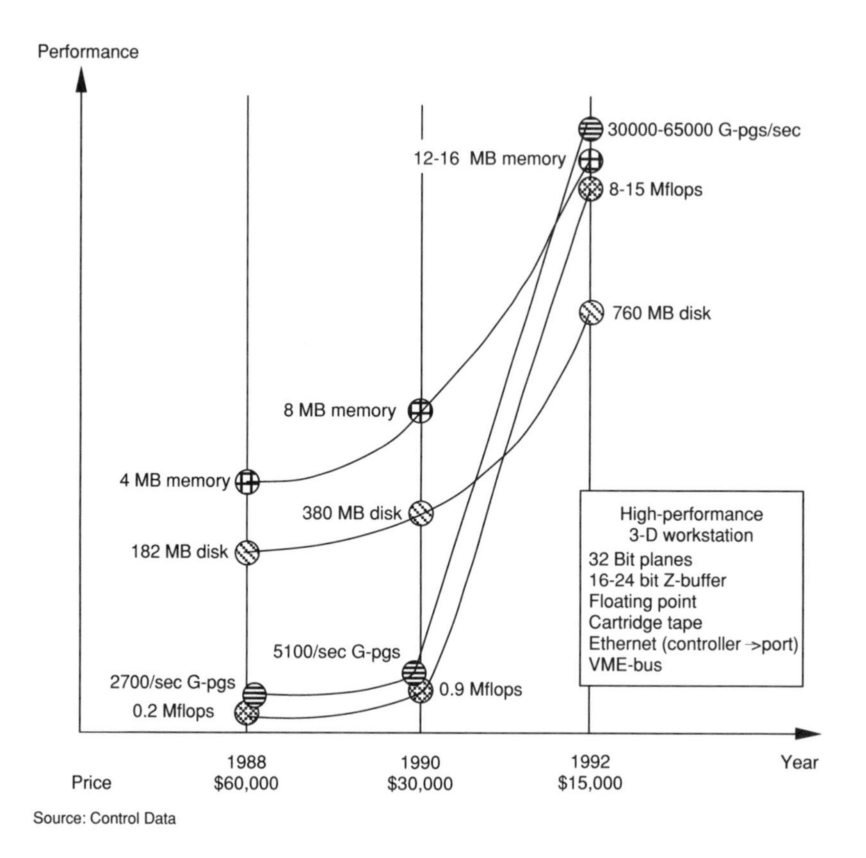

Figure 11-1. The rapid progress in the power of workstations puts CAD in the reach of even small companies.

Further improvements in performance can be gained with the use of parallel processors, such as the Transputer, acting as graphics engines, but the trend is toward the embedding of some manipulation programs in firmware in the computer. With this approach, the number of instructions required to move data will be reduced drastically.

Expert Systems

Ultimately, manufacturers need to use the same basic surface data in all departments without manual intervention. In addition, expert systems containing data such as the permissible wall thicknesses, draft angles, corner radii, and so on for castings; practical limits on deep drawing and wrinkling and deformation of metal in stampings and forgings; and data on other processes will become an integral part of engineering. These standards will be input by the specialists from manufacturing, and design engineers will be able to apply these to their designs to eliminate some of the impracticalities. However, such techniques will not replace the team approach but optimize designs. Of course, these expert systems will need to be simple enough to be updated by the company's experts on processes.

Several rule-based systems are already in use. For example, there is a proprietary program for simulating injection molding called Moldflo, and a number of companies including Computervision have developed rule-based systems for the automatic layout of printed circuit boards.

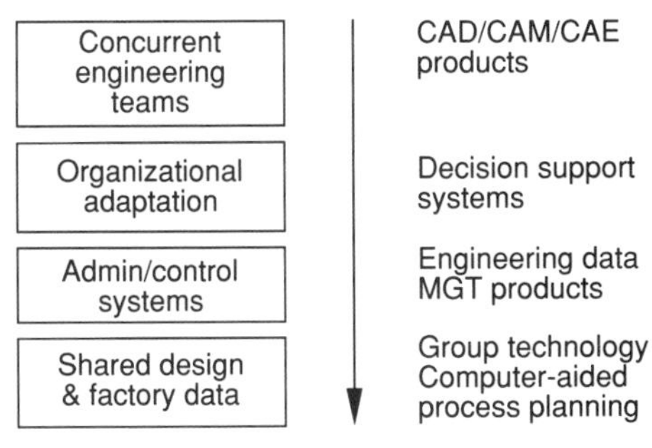

Source: ISATA, Prime Computer

Figure 11-2. CAD/CAM products match the various stages of concurrent engineering.

Simulation in Parallel

With CAD, once some rough concepts have been produced, simulation is used to test alternatives. First, the stylists can generate a number of alternatives for viewing by management. With over-the-fence engineering, everyone else would have to wait until the clay model or prototype had been built before they could begin work. Not any more; with CE and the correct approach to data handling, other engineers can begin their work before the design has full approval. Initially, they are looking at the package in terms of space for mechanical units and people to be carried, if appropriate, and then overall pattern of stress and other factors. As the design becomes hardened up, they can gradually narrow their approach.

All vehicle manufacturers should now be in a position to simulate much of the design at an early stage — either in-house or by using the facilities offered by consultants such as HW Structures, the arm of Hawtal Whiting that specializes in this aspect of engineering. HW Structures has developed a "coarse simulation" approach to enable customers to gain ballpark figures on stress levels and functions such as barrier test performance and aerodynamics in a short space of time. Thus, the customer can see the potential for three or four alternatives to a new design without producing anything in metal. That must be the target. The whole principle of CE is that as much development as practical is pushed forward from prototype to preconcept and concept stage. In moving toward this situation, manufacturers need to gain confidence that the simulations are accurate, so a progressive approach is desirable.

Three-Dimensional CAD

There are many CAD systems, such as AUTOCAD, CADDS and MEDUSA, CADAM, CATIA, CGS, Control Data's ICEM, INTERGRAPH, and PDGS. PDGS is the system developed by

Ford and Prime Computer. It is now marketed by Computervision and is used exclusively by Ford and its vendors. CAD systems are based either on solid modeling, which is used mainly by the automobile, aerospace, and consumer electronics industries, or two-dimensional systems, of which the more advanced can also build up three-dimensional models. These are used for plant layout, process industries, and in specialized form, electronics manufacturers.

Because of their ability to generate and handle complex shapes, 3-D CAD systems are highly complex, generally building up models from standard shapes, such as cubes and cylinders, that are modified and assembled. One problem with this approach is that in building up a suitable shape, the designer may alter the design many times in different areas, and then need to modify a section some time later. If the integrity of the system is to be maintained, a hierarchical system or at least a recorded trace of the designer's actions is needed. CADDS is one system that is based on a hierarchy in which the sequence of building blocks is known, so access to a previous block is simple. Also, the nurbs (non-uniform rational B-splines) surface design tools incorporated in CADDS and other packages are designed to speed up the generation of complex forms such as airfoils and automobile body shapes.

Where complex assemblies are being built up, it is necessary to establish the relationship between them in a hierarchical manner, so that even when the engine and transmission are no more than a set of external dimensions, their position and relationship with other components can be defined. This feature is particularly important with CE since it allows all departments to work in parallel, with the later versions being added to the assembly as they are developed.

Recently, the trend has been toward further development of modules for special applications and open systems that allow

CAD Wire Frame

Color Rendering

Finished Product

Figure 11-3. With 3-D CAD systems, models can be built up as wireframes and then with shading to give a realistic image.

proprietary packages to be run in conjunction with the basic CAD system. For example, Control Data, working with Volkswagen, has developed a group of simulation packages that make it possible for styling and engineering studies to be carried out in detail without any metal being cut. The basic approach adopted by Control Data for the software is heuristic, taking into account the geometry with its discontinuities. It supports automatically the iteration process typical of CE; Control Data preferred this approach to the use of the mathematically ideal geometry without discontinuities. The packages are part of a suite of programs called ICEM — integrated computer-aided engineering and manufacture — and consist of

- ALIAS, for styling
- VWSURF, for surface modeling
- DDN/3, for detailing of surface modeling

ALIAS not only allows three-dimensional views to be shown with highlights according to the angle of the light, and with the appearance of windows shown accurately, but it also includes a module that allows videos showing models in a natural environment to be produced — almost as the real car will appear in due course. This is an extremely valuable extension to the software.

Among the simulation packages are

- structural analysis
- linear static and dynamic analysis
- nonlinear static and dynamic analysis
- sheet metal forming simulation
- computational fluid dynamics
- aerodynamic tests

A common data format is used, and with these programs, many specific simulations, mainly those required by the automobile industry, can be carried out. These include

- finite element analysis
- a variety of crash tests, showing behavior of both the structure and passengers
- acoustics analysis
- torsional strength analysis
- vibration analysis
- combustion analysis
- duct design
- design optimization

Since many of these simulations can be run simultaneously by different specialists, the opportunity exists to investigate a large number of variations on a theme, and the results studied by the task force well before the design is firmed up. An idea of the benefits of the system can be gained from the way in which Volkswagen engineers optimized an engine subframe. After the initial design, the assembly weighed 50 kilograms. The CAD system iterated alternatives, using finite analysis to evaluate each, and eventually a weight of 30 kilograms and equivalent stiffness were obtained.

In the United States, one of the Big Three used CAD and CE to optimize a similar frame, but in that case the result was a frame made from one stamping instead of from three. Despite the use of CAD, it is extremely unlikely that a one-piece frame would have resulted without a team approach because the loads involved required considerable knowledge of acceptable stress levels, while the complicated shape required for the frame could not be achieved without the expert knowledge of a stamping specialist.

All development work was done at the terminal, and when the prototypes were found to deflect a little more than expected, further refinements were made to the form and thickness — with simulation, which proved more effective and faster than modifying prototypes.

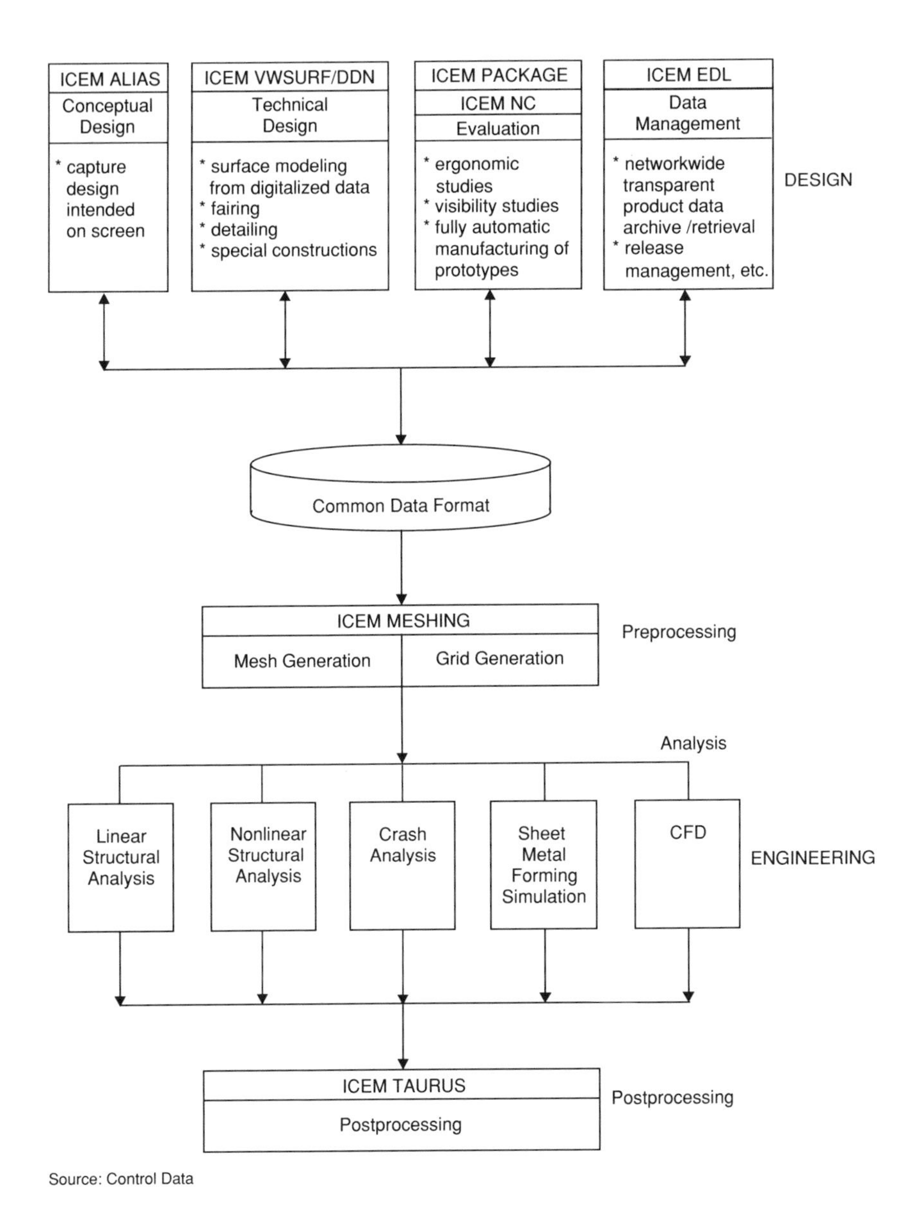

Source: Control Data

Figure 11-4. The large suite of design and simulation programs in the ICEM family.

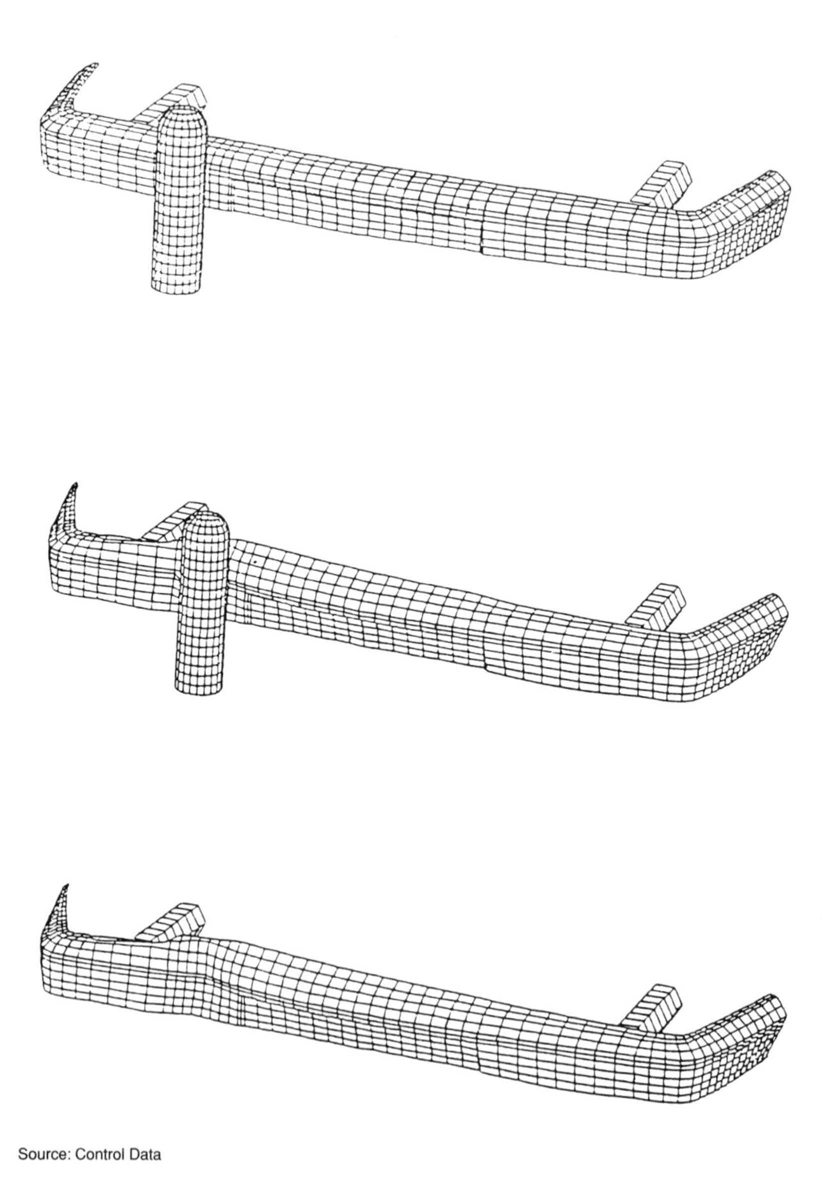

Source: Control Data

Figure 11-5. Typical of advanced simulation is a package used by Volkswagen, which not only simulates impact but shows the damage graphically.

Modules Integrated, Interface to Packages

Computervision's CADDS 4X software has capabilities similar to ICEM, with many analysis tools incorporated in the system. In addition to the solid modeling mentioned earlier (with interactive shading), there are kinematic functions to assess problems of adjacent moving parts — as with the retraction of a landing gear on an airplane or the movement of a wheel on a car in its wheelarch. Also incorporated are

- StressLab finite element modeling and analysis module
- Thermalab thermal stress analysis
- assembly design
- mapping
- interference analysis
- piping design

Another feature of importance in some applications is a capability for analog to digital and digital to analog analysis and combined simultaneous analog and digital analysis. There is also a CNC cutter path module, CVNC.

MEDUSA, Computervision's other main CAD package, interfaces directly with the GNC CNC package, and with industry standards such as PDA Corporation's PATRAN and Structural Dynamics Research Corporation's SUPERTAB finite element packages.

CAD Cuts Program Time

Among the companies using CADDS 4X in stress work and design is Toro (Minneapolis, Minnesota), which manufactures heavy-duty and domestic lawn mowers. CADDS 4X is used in the concept stage, allowing a number of "what if" situations to be considered. Once the basic design has been outlined, structural analyses are conducted using StressLab finite element modeling software.

After approval of the design, the same model data are used as a basis for CNC machining of prototypes, with the tool paths being created by CVNC software. At this stage, the cutting of the metal can be simulated on screen, so that many glitches can be eliminated before the data are postprocessed into the appropriate language for the CNC machine.

With the aid of CAD and the CE methods it brings, Toro has been able to speed up its model introduction programs. For example, the RN223-D diesel mower designed to cut grass on the fairways of golf courses was taken from initial design to production in 22 months, instead of more than 36 months for the previous model.

Wiring Design by Rule

Rule-based systems are a key to reducing the time required to develop new products, but only if they result in sound designs. One new module of this type that has already proved its worth is Computervision's Harness Design. It is based on software designed for piping layout, but it has been extended substantially to allow the design of wiring harnesses for vehicles, airplanes, and similar assemblies. As a first step, the designer completes a schematic layout of the vehicle; in practice, by the time the wiring harness is ready to be designed, the layout will already be available in the files. Next, the designer adds all the electrical components in block form, and includes data on the loading of the units. Finally, he or she adds approved channels for the wiring.

With an CE task force involved in the design, the location of the channels would result from input from manufacture and service as well as design. Once this has been completed, Harness Design lays out the wiring harness, specifying appropriate sizes of wire. Finally, the three-dimensional harness is converted to a two-dimensional shape to produce dimensional data for the production of the manufacturing fixture. In one

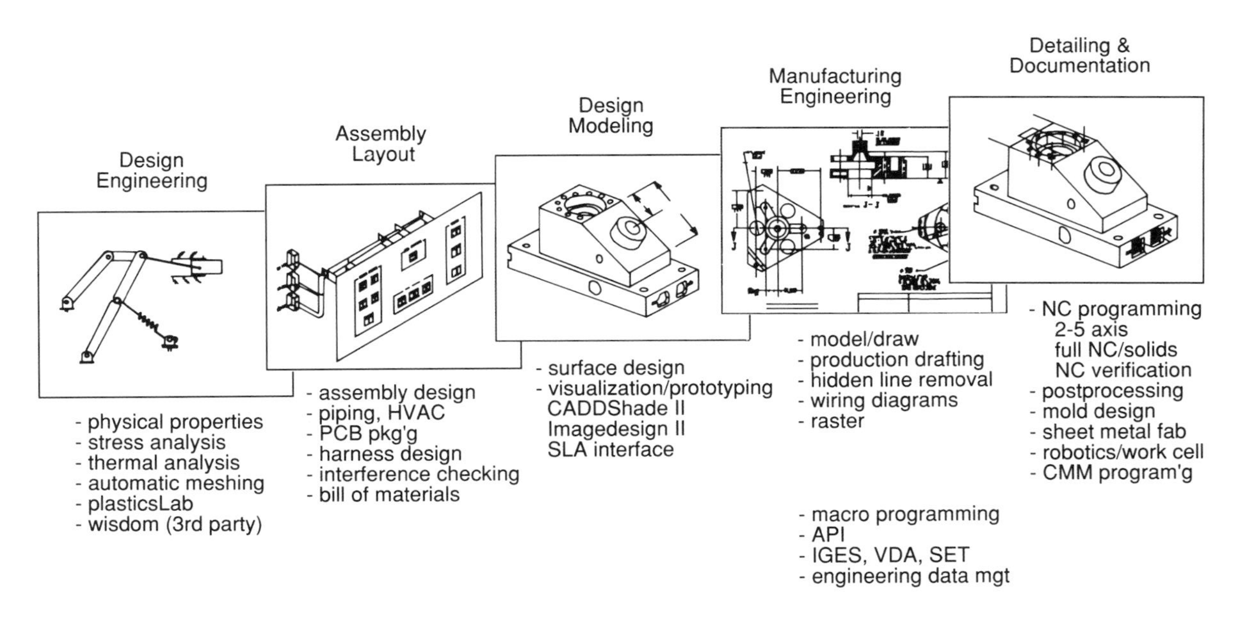

Figure 11-6. The applications available in CADDS 4X extend from design to manufacture.

case, the use of Harness Design on a vehicle reduced the lead time from 16 to 4 months. Boeing, which uses IBM CATIA for its mechanical engineering design, adopted CADDStation to handle wiring diagram and schematic work.

Automating PCB Wiring

There are many approaches to the automation of laying out printed circuit boards (PCBs) and completing the wiring, but many result in a poor design that needs to be refined substantially by the designer. Computervision's THEDA is an advance on most with its repositioning router. Users specify rules for clearances around ICs, for example, and can also build up a library of standard units such as memory chips, resistors, and capacitors. They then produce a logical layout, and indicate the space available on the PCB. If they wish they can locate certain key components, or they can leave the software program to do the complete layout.

When it comes to the wiring in the PCB, the normal process adopted by CAD systems is for each extra operation to be built on to the existing layout, literally going around problems rather than solving them. However, THEDA does not do this, but repositions wires as necessary when it adds new ones. For example, as it builds up the wiring layout, it will come to areas where one wire is already crossing another at right angles. At the crossover point, one wire is taken through the PCB and taken back up after it has crossed the other wire.

Because of this crossover, there is no space to lay another wire straight across the PCB, and so a conventional system will take the wire underneath, complicating manufacture and weakening the PCB with four holes very close together. By contrast, THEDA moves the holes further apart and lays the wire on top of the PCB. Also, if necessary, it can move several parallel wires across to leave space for another.

Because of the relative simplicity and controlled situation in the manufacture of PCBs, THEDA also takes into account a number of rules related to manufacturing, which is still the weak point of CAD systems.

Japanese Make Tools from CAD Data

Japanese automobile manufacturers have tended to develop their own CAD systems, either from the outset or from a standard package, because they require a larger number of functions than American companies. One reason for this approach is that from the outset, the Japanese have been using some aspects of CE, with integrated design and manufacture. For example, both Nissan and Toyota have their own systems.

In the Japanese automobile industry, the target in the late 1970s was for the production of dies for body stampings to be produced directly from CAD data, and the Japanese are ahead of their American rivals in this respect. Toyota's CAD system extends through the manufacturing system, from styling to the manufacture of dies and welding fixtures.* As a start, equations that permit curves and highlines to be expressed simply are used by stylists. The resulting data are used to control a CNC machining center producing the clay model — normally produced by hand. Any changes required are input at the terminals to maintain the data base. Subsequently, the data are used by body designers to design actual panels. Although Toyota found that large body panels could be produced three times as quickly as manually, a gain of only 30 to 40 percent has been made so far in the design of small panels.

Another software package, DIEFACE CAD, has been developed for the computerized drawing of die faces. Depth of draw

* M. Ohara, "CAD/CAM at Toyota Motor Corporation," *Japan Annual Reviews in Electronics, Computers & Telecommunications* 18 (no. 3:2).

is shown on the screen, and as a result of the success of this package, the use of plaster models for die design investigations has been eliminated completely.

A master data base is used to store dimensional data so that tools for body pressings can be generated without the need for a master model. Together with some refinements in machining, this system has cut lead times significantly. The system has also been extended to generate data for welding and aperture checking fixtures, which were previously built from master models.

Common Data Base Needed

Toyota's system represents a significant step forward, and as mentioned earlier, the target is for a common data base to be available to all departments in the form they require, and for automated processing of these data as required. The data should be available as

- design data for product engineering and component suppliers
- functional design specifications for specialist suppliers
- manufacturing data for manufacturing engineers
- full specifications for cost analysis
- specifications in product terms for marketing

With such a system, CE will bring even greater gains than at present. In the meantime, there is a need for an integrated method of storing, backing up, and protecting the data base. Computervision's EDM is one solution; it stores, catalogs, and retrieves engineering data from any sources in the company. Sources can include process plans, NC programs, and noncomputer generated data. One important feature is the automatic backing up of data.

The full integration of CAD and CAM data are a long way off, however. As mentioned, many programs allow the use of

CAD data to be converted to CNC data. Software for two- to five-axis CNC machines is available, and Computervision is among those whose software can be used to simulate processes on machines such as punch presses, lathes, laser cutters, and water jet cutting machines. Collision detection is incorporated in the Computervision programs.

Emphasis on CAM

CIMLINC is one CAD supplier that has concentrated on manufacture as much as on design, with the development of some special packages and a system for the control of DNC machining. Moreover, it has developed translation software so that data from most of the popular CAD packages, such as CADDS and CADAM, can be run directly by its programs. It claims that its CIM CUT program is much quicker than others in generating CNC programs, citing 28 hours work from surface data against 140 hours with conventional CAD/CAM methods, and 52 hours against 160 hours from 3-D wire frame data. Another package from CIMLINC, CIM CMM, has been designed specially to translate CAD data into a form suitable for a coordinate measuring machine. This makes it possible to measure prototypes and production samples quickly. Computervision's CADDStation system has also proved its worth in the design and production of dies. Greenville Tool and Die (Greenville, Michigan) has been using Computervision systems for a decade, and recently upgraded to CADDStation. Among Greenville's customers is Ford, so it receives surface or wireframe model data in PDGS format, and can convert it directly to CADDS. Then, running CVNC, it can create tool paths much more quickly than previously — 90 minutes against 18 hours in some cases.

CIMLINC's ID package, which is used to produce process sheets and other documents that combine written instructions and drawings, has also proved a success, partly because it can process data from other vendors' software and partly because electronic documentation is at the heart of DoD's CALS.

Martin Marietta, the aerospace company, and Harley Davidson are among the companies that use ID to produce process sheets. Harley Davidson finds that in a typical case, its use has reduced the time required from 10 to 2 days.

Despite the progress made by Toyota, few manufacturers have much confidence in the use of automated systems to generate dies for stampings or molds for die castings. The problem is that in the stamping of metal deforms, nonlinear processing, which requires enormous computing power, is required. Meanwhile, the behavior of metals as they cool and solidify is not fully understood. On the other hand, Moldflo, which incorporates much empirical data on injection molding, is a successful package for the development of mold shapes. For other processes to be handled in the same way, more power and more knowledge are needed.

Small Manufacturers

But what of the small manufacturer or vendor? How can it take advantage of this technology if the software requires such powerful computers and high-resolution graphics? Clearly, the existing systems are too expensive for all designers to be able to use CAD, but with workstations now replacing minicomputers, the prices of CAD hardware/software systems have fallen, and this trend is accelerating.

Even so, a fully fledged CAD workstation with extensive software costs around $60,000, and a typical installation of four workstations would cost around $150,000. On the other hand, some of the more powerful PC/workstations are now able to run CAD programs effectively, so it is possible for a small company to do its own work for $10,000 to $20,000, depending on its choice of equipment. When the company expands, the linking of workstations with a server and network allows inexpensive expansion. Where the company is a vendor to several companies, it will need to be able to translate data from its customers' system into its own, a factor that will influence the selection of hardware and software.

These companies are faced with difficult decisions, but prudent and imaginative investment in CAD equipment will gain them a competitive edge. In any case, the use of CAD/CAM systems is inevitable for companies wishing to achieve world standards in lead time and quality. Their use enhances the potential of concurrent engineering in achieving these goals.

CHAPTER TWELVE

Concurrent Engineering Makes Friends with Manufacturing

Production problems eliminated at source
Opportunity to test new manufacturing methods
Manufacturing feasibility at concept stage

Concurrent engineering is the key to exploiting the full talents of a company's manufacturing engineers. With over-the-fence engineering, manufacturing engineers are fighting with one hand tied behind their backs — much of their knowledge and experience cannot be used because they are given details of the product too late. Instead of having time to test new ideas, they have to use methods that are already proven. Yet the product designers have been able to consider new concepts, however impractical they may be.

Unfortunately, there is a natural antipathy between product engineering and manufacturing engineering — and with purchasing for that matter. Product designers are apt to return from meetings with manufacturing engineers talking of "frustrated designers" trying to change the design. They tend to reject out of hand any suggestions to improve the design from outside their department, unless the changes are essential to make production feasible.

In any case, by the time the manufacturing engineers suggest that changes be made to the design, it is usually too late for them to incorporate any new ideas they may have. Either the design has been frozen already or there is insufficient time or funds to build a test cell to prove the new methods — and that is just as bad as committing a company to a design that has not been fully specified.

Over-the-Fence Design

Examples of designs that are difficult to make being thrown "over the fence" are commonplace. In one classic case, a multinational automaker devised a competition among design teams in three different countries to produce the best cylinder head. "Best" meant a head giving good overall performance, fuel consumption, and emissions. The U.S. team produced the best design, and this was then thrown over the fence for manufacture in several different countries.

Although the performance of that cylinder head has proved satisfactory, it has been a problem for manufacturing. In the center of the head, running from front to rear, is a very thick mass of aluminum that is subsequently drilled and bored out. That mass of metal led to problems of porosity. Secondly, since the casting is very complex, the machining operations are costly, leading to many rejects being discovered after machining, compounding the real cost. The design would have been far easier to make had manufacturing engineers been involved in a project task force from the outset.

Castings, particularly those used for engine cylinder blocks and heads and for transmission casings, are highly complex in form, yet the designer is interested only in certain aspects and features of the component. In some areas, so long as strength is adequate and weight is not increased, the precise form is not important to the designer. However, many areas are of great importance to the foundry, so these components should always be designed with a foundry specialist on the team.

Two-Part Solution

Contrast that with the situation involving a task force in which the foundry specialist on the team and the manufacturing engineer, unhappy with the prospects of the difficult casting and complex machine, request a rethink on the design. After a brainstorming session, the task force decides to try a new concept: a head cast as two pieces, split horizontally through the water jacket so that the halves can be pressure die cast and then bonded together.

The use of pressure die casting eliminates the need for sand cores and results in a much shorter cycle time, so that tooling costs are not excessive. The castings can be inspected for porosity in a simple way, and since the mass of metal is now cored out, wall thicknesses are uniform, so the likelihood of porosity is reduced anyway. Because dimensions of the casting are close to finished size, machining costs are reduced as well. The precision of the casting is expected to result in improved port and combustion chamber shapes, with the prospect of an improvement in performance.

In theory, then, the task force has converted a problem for the foundry and machine shop into an easy task. Since the two halves of the head never need to be separated, there are no complications in service, while durability is assured by the compression forces applied by the head-retaining bolts. Nevertheless, this is an unproven idea, although one that a number of European manufacturers experimented with in the 1960s and 1970s.

Because the idea is accepted as a possibility from the outset, design and manufacturing both have time to undertake the rigorous testing needed for such a new concept, with the potential of turning a troublesome design into a simple and inexpensive one. More testing than normal is required on rigs and on the road. Whereas that would probably be impractical with over-the-fence engineering, it can be accommodated easily with CE.

At the same time, as a backup, tests can be made with a conventional one-piece head, modified in an attempt to eliminate the worst problems of the design. In that way, the team is bound to finish the program with a better proposition than they started with.

Moreover, since the purchasing officer is part of the task force, he or she knows in detail what type of adhesive is required from the start. Quite simply, once a task force is adopted, radical design, whether it be to improve performance or ease of manufacture, has a chance, and manufacturing can push its development work through in parallel with product design. Indeed, both product designer and manufacturing engineer are involved in all aspects of the work.

Product Designer in Charge

One leading manufacturer, before it embraced CE principles, had a system whereby manufacturing had to accept a design by signing it off before assuming responsibility for production. This system is not so different from that used in the automobile industry for years. It was thought that this approach would prevent product designers from producing assemblies that could not be made easily.

But this proved not to be the case. Since product designers had the ultimate authority for design, they would often reject suggestions from manufacturing, so that when the device went into production, manufacturing would refuse to accept the product. Therefore, the chief engineers found themselves responsible for units in production, and whenever a production problem was encountered, the manufacturing staff would come to them to find out what to do. This unsatisfactory situation was changed once the task force approach was adopted.

There is also the matter of flexibility in manufacture. Today all product design and manufacturing engineers recognize the need for flexible manufacture, but they have different ideas on what they want and what a flexible system can achieve. In the automobile industry, the accent is on the ability to run more than one model down the same line, and to rebody a car with the minimum change in tooling. In machine tool manufacture, it means making the maximum number of models on the same set of machines — perhaps with an FMS making spindles for all machines, and some machining centers doing all the work on machine beds.

Clearly, it is important that both product designers and manufacturing engineers understand the limitations of a flexible line for machining or assembly, and that manufacturing engineers know exactly what the designers might want to do halfway through the life of the machinery. With conventional engineering, they do not work close enough to gain that understanding; with CE, they do.

Accentuate Adding Value

So, are the benefits of CE to manufacturing purely the elimination of some bad practices? Not at all. CE releases the manufacturing engineers from the straitjacket of a rigid design and gives them freedom to contribute fully to the project, adding value all the time. They change from being passive members of the company to active members of a team, able to bring knowledge of new technology and improvements in the plants.

By being part of the team, manufacturing engineers contribute to an improved design, but they also stand to increase the efficiency of their own activities. A study within Lucas Automotive showed that only a small amount of the work carried out in the process of planning, ordering, and setting up manufacturing plant was adding value. The proportions of the main operations adding value were

- plant ordering, 35 percent
- drawing manufacturing layout, 31 percent

- ordering drawings, 39 percent
- ordering equipment, 32 percent
- requisitioning tooling, 32 percent

Generally, two-thirds of the time spent by people involved in this aspect of the project does not add value. Some is spent waiting for information — put at 10 to 15 percent by an engine manufacturer — and some of it is spent on work made redundant by decisions taken later on. To increase the proportion of added value in these operations, manufacturing engineers must be involved in the task force.

Manufacturing engineers should be in the team from the original concept stage, gaining data on customers' requirements with other team members and discussing the direction in which the concept should go. So, after a very short time they have an in-depth knowledge of the volumes, the number of variations, and the general concept of the design.

Of course, changes will be made, but the engineers will not take shocks back to their department, such as the need to change completely some dies for body panels or the pitch circle on a transfer machine, a few weeks before production is due to start. In the early stages of the project, they will take back different ideas every week, but the detail will not be needed. This does present other problems, in that the manufacturing engineering department has to adapt to the idea of working on concept drawings, then with incomplete drawings, and finally with complete drawings where product design gives them confidence that no major changes will be made, and later that no minor changes will be made.

New Manufacturing Technology

The task force is far more likely to be able to keep up to date with changes in manufacturing technology than with sequential engineering. For example, they need to know whether they can use sinter-forged connecting rods, fine-blanked gears, and whether laser welding already coming into use for automotive sheet metal is a practical proposition. And can they emulate Saturn and use plastic body panels?

The best source of knowledge on the practicality of new processes is the manufacturing engineers — because they know their plants and will have the job of planning the equipment. However, without the task force approach, they will not be aware of why the designers want to use a new technique or what it does for the product, and so they may dismiss it as impractical. Once in the team, they will be filled with the same enthusiasm as the designers, and so will endeavor to find suitable new technology, but only if it is feasible.

Manufacturing engineers therefore gain both enthusiasm and knowledge by being in the team. Even if the processes are routine, they will be able to build up their ideas of the production equipment as they go along. For example, the plan may be for 2.4 and 3.0 liter versions of a new engine. However, it may become clear that there is demand for a diesel, which would need to be 3.5 liters in capacity. Later, designers may wish to consider rationalizing the design so that another unit of 2.0 liters can be replaced as well. Because they are involved in all the discussions, the manufacturing engineers will be able to contribute in a positive way.

It may be that the 2.0 liter engine is a popular unit, and that apart from the cylinder head, the design is still reasonably modern. The design engineers might assume that because the plant is fairly old it will be cheaper to adapt the new cylinder block to the smaller unit, even though the 2.0 liter unit would then be heavier.

Inside Knowledge

The manufacturing engineers will know the line is actually running efficiently and that there is plenty of life in it yet.

Therefore, the team might decide to retain the old 2.0 liter block but design a new cylinder head, to be built on the new line. A side effect of that decision would be that the volumes for cylinder heads are increased sufficiently to make an automated line for valve insertion economic. And so it goes on.

Also, the manufacturing engineers can decide more easily whether to plan the line at the outset to cater for all derivatives, or simply leave space for additional machines to be installed when later versions are added. They can apply the same principle to the decision on automation, knowing the full range of derivatives, volume, and timing. There can be no shocks about the meaning of flexibility — the product and manufacturing engineers can discuss it continually.

In that type of project, continuity, coupled with the fact that the manufacturing engineers have time to test new ideas at the concept stage — such as automatic assembly of the valve gear for a four-valve version — are vital. Not only can the manufacturing engineers carry out their job better, but their increased involvement in the project raises their morale.

Reduced Manufacturing Costs with CE

There are many cases where the involvement of manufacturing in CE teams has cut costs. There is the replacement of a one-piece housing of an automatic transmission by three casings cited in Chapter 4. In another example, product design proposed a cylinder head with conventional rocker shafts and rockers. The manufacturing engineers felt that the design was too complex. They suggested instead that pressed steel rockers be mounted on pedestals. The elimination of the shafts and bushings reduced cost and simplified production, all because manufacturing engineers were involved in the project from the start.

In another case, a spring, clip, and bushing of a pedal box assembly were redesigned as a result of tests by manufacturing engineers at an early stage. The components were redesigned for automated assembly, but as is so often the case, they are easier to assemble manually as well.

In designing a new housing for its minicomputers, the task force at Digital Equipment gave manufacturing functional requirements. After due investigations it was proposed that the nuts and bolts be replaced with snap fasteners — 50 screws were eliminated. A new coating for the steel panels was also proposed. The result was a cost savings of more than 50 percent.

Radical Method

More dramatic was the improvement made by GM at its Adam Opel subsidiary in the design of a rear suspension arm. In order to exploit CE, Opel set up a new Advanced Product Study (APS) department consisting of both product design and manufacturing engineers. They work together on the design, with the manufacturing engineers assessing feasibility and costs. Although the product designers use CAD stations, the production engineers do not; it is considered better for product designers and production engineers to sit at a desk or at a terminal to discuss feasibility. Therefore, confrontation is replaced by teamwork.

To replace a heavy cast iron arm, the design engineers proposed an arm stamped from two halves that were subsequently welded together. The new design would be much lighter than the old one, and machining would be reduced. Nevertheless, the manufacturing engineers did not like the design, for a number of reasons:

- The shapes were such that despite optimizing the nesting of the blanks, 39 percent of the blank was scrap.
- Since the two arms were not symmetrical, four sets of dies were needed.

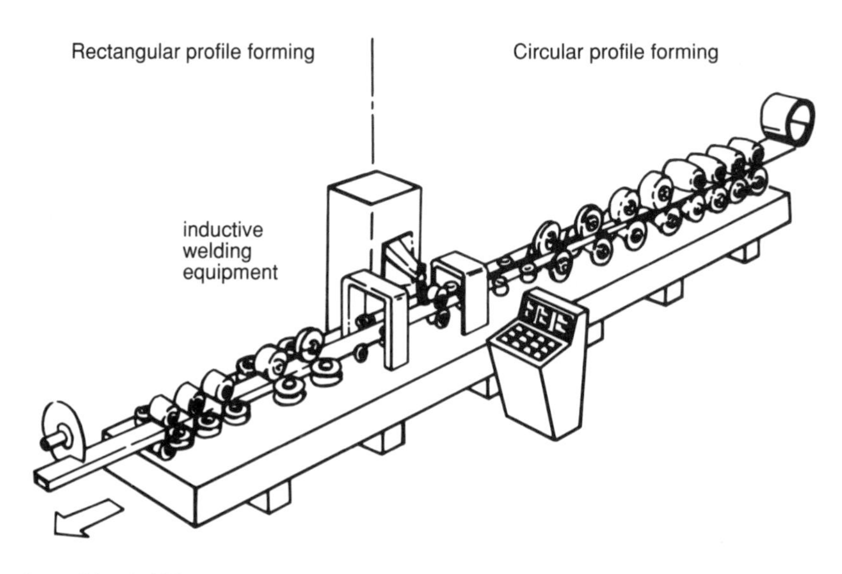

Source: Adam Opel AG

Figure 12-1. The main sections of the Opel suspension arms are formed on a standard tube mill.

- A CO_2 welding seam, 1.33 meters long was needed on each arm — a very costly process.

The product and manufacturing engineers therefore set about designing a new arm. Manufacturing suggested that the arm be built up from rectangular-section tube with a fabricated hub carrier. The arm was redesigned on these lines, with the result that the members of the arm could be produced on a tube mill — a standard piece of equipment suitable for high volume production. The mill converts sheet into a circular tube, which is welded automatically. Then, the tube is formed into rectangular section on the same tube mill. Circular tube produced on the same machine, with the same tooling is used for the rear suspension cross member.

Another standard machine, a CNC pipe bender equipped with a ceramic mandrel, is used to bend the tube to the required shape. Subsequently, the tubes are drilled, cut to length, and assembled. Only 245 millimeters of weld length is required, compared with 1,330 millimeters for the previous design. The gains made with this design over the original one are

- lead time cut from 14 months to 3 months
- weight saving/arm, 5 kilograms
- weight saving/cross member, 15 kilograms
- scrap reduction, 5.4 kilograms
- cost reduction, 9 percent

In addition, the thickness of the tube can be altered to suit the load capacity — the internal dimensions of the tube do not affect tooling — so it is suitable for different models.

Such a savings would not be practical without the use of concurrent engineering. Instead, the type of changes that would have been practical would have been to make the stampings symmetrical to reduce tooling costs. Other changes would be out of the question, partly because the design would have proceeded too far and partly because the changes would have increased lead times too much.

A manufacturer of electric hand tools also made worthwhile reductions in costs with its first CE program.* This was a pilot program, involving the optimization of a gear drive for a hammer drill only. In the first approach, a helical gear with an integral hub was designed. Since the bore of the hub incorporated drive serrations, three machines were required to produce the gear:

- A lathe to turn the blank to the basic form

* W. Ritter, "Simultaneous Engineering: An Organisational Prerequisite for Efficient and Rapid Technology," in *Proceedings of the 1st International Conference on Simultaneous Engineering*, London, December 1990, pp. 53-63.

- A broach to form the bore
- A gear hobbing machine for the gear teeth

After a brainstorming session at which the cost of manufacture was highlighted, a new design made from two parts — the gear and a separate internal ring — was evolved. So that the two components could be produced from one fine blanking machine, spur gears of suitable strength were designed. The two parts were then welded together. The use of two parts instead of one goes against some of the tenets of DFMA, but the use of an innovative production method overcame the potential disadvantage.

Because the one-piece gear was of complex shape and was machined from the solid, 80 percent of the material was machined away and lost as swarf. With the two-piece design,

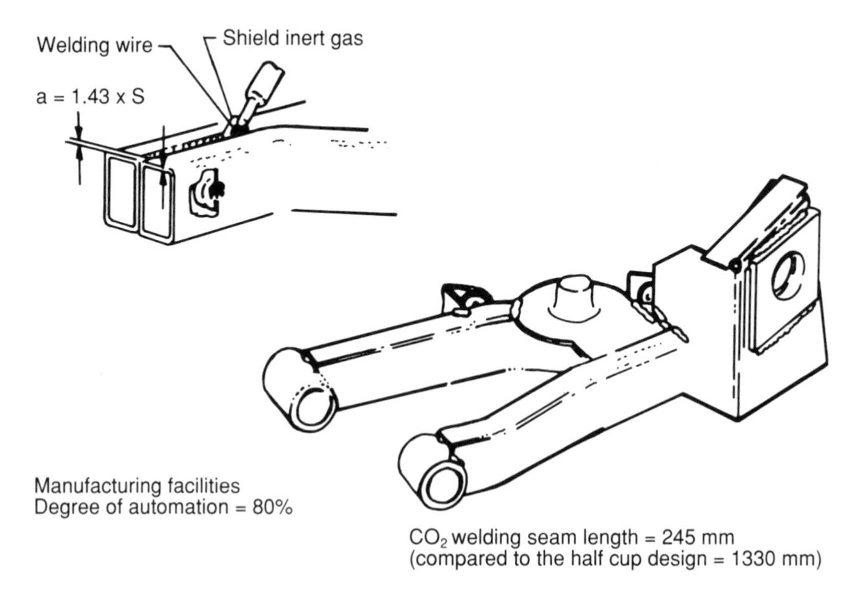

Figure 12-2. The CO_2 welds in the new Opel suspension arm are much shorter than in the original design.

only 50 percent was lost as scrap, which is still high. However, because fine-blanking is a fast process, whereas gear hobbing is very slow, the two-piece gear cost 40 percent less to produce than the earlier design.

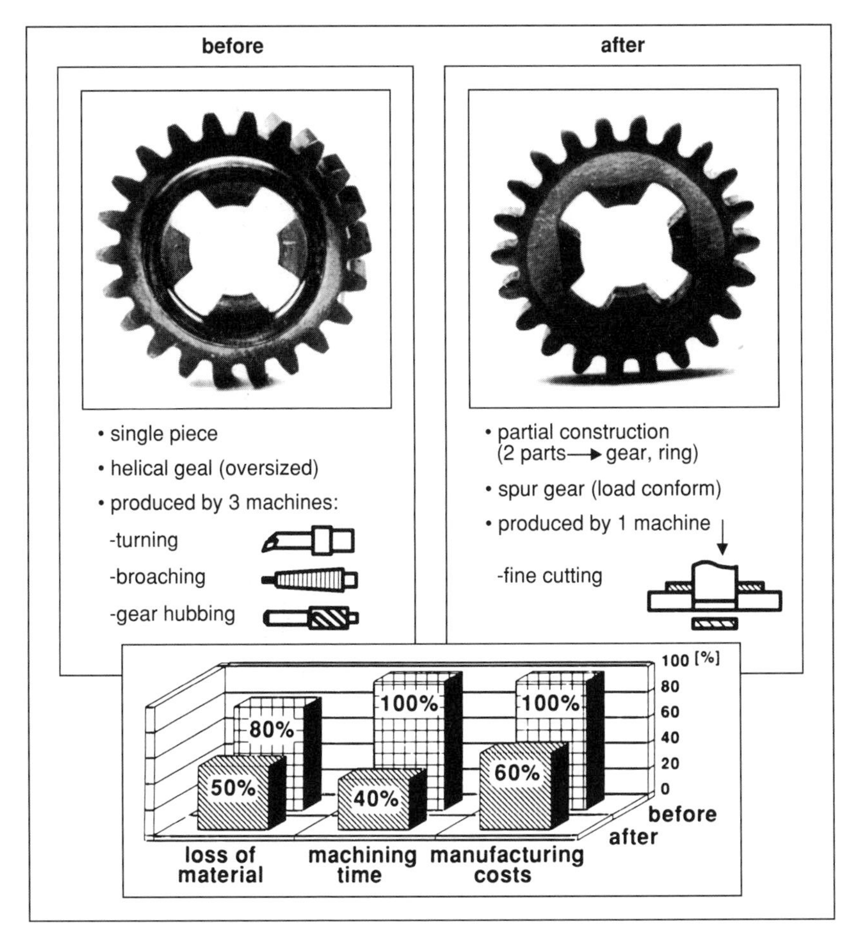

Source: I.W. Eversheim, "Trends and Experience in Using Simultaneous Engineering," *Proceedings of the 1st International Conference on Simultaneous Engineering*, London, December 1990, p.20.

Figure 12-3. The replacement of a one-piece gear with a two-piece design cut costs by 40 percent.

Tools to Ease Manufacture

Several methodologies can aid the task force in reducing manufacturing costs. For example, Taguchi is just as useful in refining a process in manufacture as it is in eliminating excessive scrap in a product. DFMA, discussed in Chapter 9, includes modules to assist in the simulation of assembly lines. Software packages suitable for simulation of forging, die casting, investment casting, metal cutting, heat treatment, and queuing in transfer lines and other conveyor systems are also on the market.

Chrysler uses simulation of its processes to assist in optimizing design and plant, and uses assembly variation analysis to simulate the effects of tolerance stack-up on the design. This technique allows "what-if" tests to be made with different tolerance levels.

Although such aids can be used in conventionally engineered programs, the whole approach lacks the urgency and immediacy engendered by a task force, of which all the members are anxious to find out whether a certain approach is feasible or not.

In designing the machinery to be used in the plants, the engineers should not ignore QFD and FMEA. These tools are just as valuable in machinery design as in product design, and the use of FMEA is now an established procedure at some machine tool builders. Manufacturing also needs to upgrade its efficiency and to think in terms of whole life costs. Simulation of hole drilling should not stop at the costs of different methods and tool life, but needs to include the cost of removal, sharpening, replacement, and the labor involved. In return for its involvement in the task force, manufacturing needs to be able to deliver across the board, and that includes the need to increase the typical uptime of a transfer line from the 50 to 60 percent standard in the United States to the 80 to 90 percent achieved in Japan.

GM Engine Division adopted CE fully throughout its 13 plants in 1989, and its first task was to implement a program to standardize the processes used in the plants.*

Because of the long life of capital plant, that will take some time to achieve, so the division has included continuous improvement projects in its plans. Since machine tools account for 75 percent of the cost of investment in engine manufacture, it has instituted various programs that involve machine tool vendors:

- Research and development into new techniques
- Program planning
- Program implementation
- Continuous improvement

With this attack on all aspects of manufacture, it intends to continually upgrade performance and take advantage of improved technology in manufacture.

Fixtures to Add Value

CE is designed to encourage a new way of thinking about a product and how it is produced. Henry Ford taught that the distance between different operations should be as short as possible because transport does not add value, yet in many plants unnecessary distances are traversed by components and assemblies. In addition, many operations do not add value. For example, a turn-over fixture, or a device to turn a component through 90 degrees, may cost as much as an assembly station and does not result in any work being done on the component.

* J. Wallace, "Simultaneous Engineering from an Engine Manufacturer's Viewpoint," in *Proceedings of the 1st International Conference on Simultaneous Engineering*, pp. 129-135.

Production engineers need to remember DFMA and ask the following questions:

- Is the device really necessary?
- Could the design be changed to eliminate it?
- If not, can the operation be carried out from beneath the assembly?
- Can all subsequent operations be done with the assembly this way up so that a second turn-over fixture is not required?

Because they are part of the task force, the production engineers will receive realistic answers to their questions. Indeed, unless they are part of the task force, any modifications will likely be too late for all the changes to be made without delaying the program.

There can be absolutely no doubt that the use of a task force that includes a manufacturing engineer speaking with authority equal to that of the other members is a vital ingredient for success with CE. If the engineer is regarded as an advisor only, his or her presence is unlikely to be useful. This is a major priority in setting up the task force, and some companies think that a task force consisting of personnel product design and manufacturing engineering is CE. It is only part of the whole, but a very important part.

CHAPTER THIRTEEN

Customer-Driven Vendors

CE cuts lead time for capital plant by 6 to 12 months
Vendors in the task force become a resource
Close liaison leads to better components

Vendors of capital equipment and key components should be as much involved in product development as manufacturing engineers, and this is one more facet of CE where the Japanese have built their own advantage. Starting some 30 years ago, they built up pyramid structures, with each manufacturer sitting on tiers of vendors. In the automobile industry there are often four or five tiers, with the bottom tier consisting of tiny workshops specializing in a few operations.

Generally, the manufacturer has an equity stake in most of the first tier of vendors, and some of these may have a stake in those below. An important aspect of the structure is that each of these vendors will make components for one automobile company only — and that will account for 60 to 90 percent of its business. The same situation prevails in the electronics industry. The result is that the manufacturer has the undivided attention of its vendor and can involve it in advanced projects with confidence. This is true even though Japanese manufacturers maintain a stricter level of secrecy about forthcoming projects than American manufacturers.

Japanese and American Groupings Differ

Traditionally, large manufacturers in the West have adopted an antagonistic attitude to their vendors; they set one against the other to achieve the best price, and juggle orders for the same component between two or three vendors. Some have been known to take the business away just when the vendor had eliminated early problems and moved into full production.

Even apart from the psychological disadvantages of this approach, few purchasing managers realized the damage this was doing to their company's costs or quality. For example, in one case, three die casting foundries were chosen to produce a complex casting; since the orders were therefore small, each production run was too short for optimization of the process. Because the purchasing manager did not understand the problems of die casting, he ended up paying extra, not just for the three sets of dies and a higher scrap rate but for 100 percent X-ray inspection as well.

Over the past five years this situation has changed, with the multiple-source policy being replaced by single sourcing. The more enlightened companies now seek long-term relationships with vendors. Single sourcing is an important prerequisite to CE; without it, it is impractical to draw the vendors into the team, and even if it were not impractical, the vendors would not have sufficient confidence in the customer to give of their utmost. A long-term relationship follows naturally.

Indeed, GM even involves the suppliers of steel sheet in the CE team* and VW insists that one of the first operations of its task forces is to decide whether to "make or buy" components, and then to involve a vendor. This is a logical approach, partic-

* D. J. Patrishkoff, *Early Sourcing and Involvement of Steel Companies on the 1989 Cadillac DeVille and Fleetwood Programs*, SAE Technical Paper 890342 (1989).

ularly now that large manufacturers are tending to buy units required in low volumes from competitors, or set up joint ventures to make them, concentrating their own plants on core activities with high-volume output.

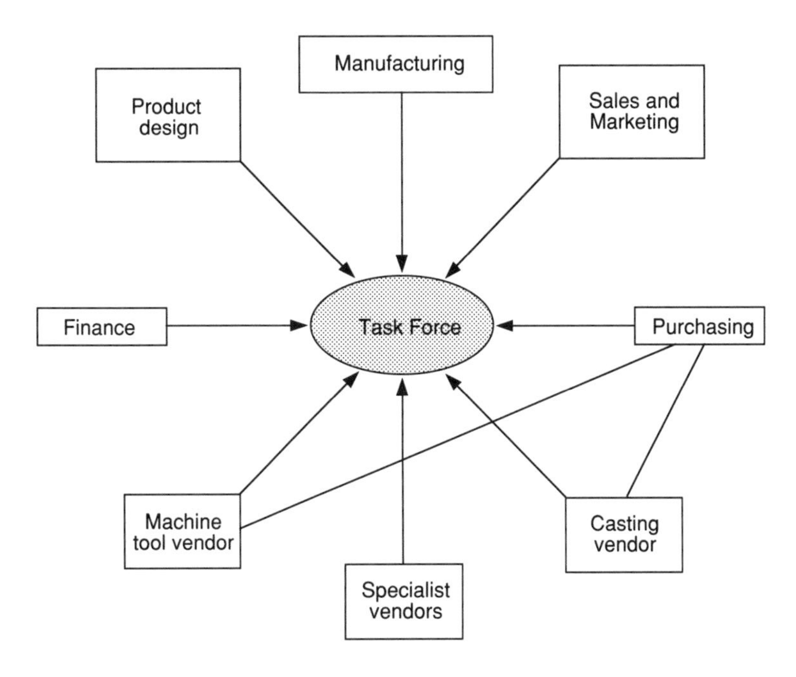

Figure 13-1. Vendors of machine tools, major castings, and specialist components should have members in the task force.

Vendors of Capital Plant

Obviously, vendors that build machinery and those that supply components are involved in the task force differently. But in both cases, it is the key vendors that need to be involved continuously, although other vendors may participate on a part-time basis.

Before bringing a maker of transfer lines onto the team, the manufacturer needs to have confidence in the firm. The approach at GM is for the manufacturing department to

- determine which vendors are appropriate for the manufacture of the component
- request five or six vendors to demonstrate their general approach to manufacture and how they would support the task force
- select one vendor before the program starts, and make it part of the team

Use the Vendor as a Key Resource

Significantly, GM Engine Division refers to the vendor as "supporting the task force." This is just as important as the quotation for the machine. In some cases, American manufacturers, including some GM divisions, ask the vendor to quote for the capital equipment for an outline product. It may be similar to the one in the project, or it may be a concept drawing. When a vendor is selected, there is normally no hard-and-fast contract; instead, a letter of intent setting out the aims of the relationship is considered adequate.

Other manufacturers prefer a more detailed contract, with the vendor giving a firm quotation. In that case, the quotation is for a certain set of machines, and it is best to agree beforehand how variations will affect the total cost. For example, the vendor might give prices for different sections, so that when changes are needed, it is known what proportion of the price could change.

The advantage to vendors of the less formal approach is that they spend less time preparing competitive quotations and have some idea of their likely work load. Generally, vendors are paid for engineering work and for the machine, so that should they complete the development work and then decide they do not wish to supply the equipment, they can be paid for their work. This happened in one case at Ford where the vendor was involved in an assembly automation project but decided that it would not be profitable to produce the chosen equipment.

However, vendors are wary of being involved in too many projects where they are paid for engineering work only. To maintain profitability, they need to generate three to six times as many work-hours in the factory as in their design department. Also, they do not want to waste a lot of time preparing detailed proposals for CE projects if they feel that they are just being used as a cost comparator for the main vendor, because a great deal of time is involved. So, although machine tool vendors want to be involved in CE projects, they have as many concerns about overinvolvement as their customers. This is one reason

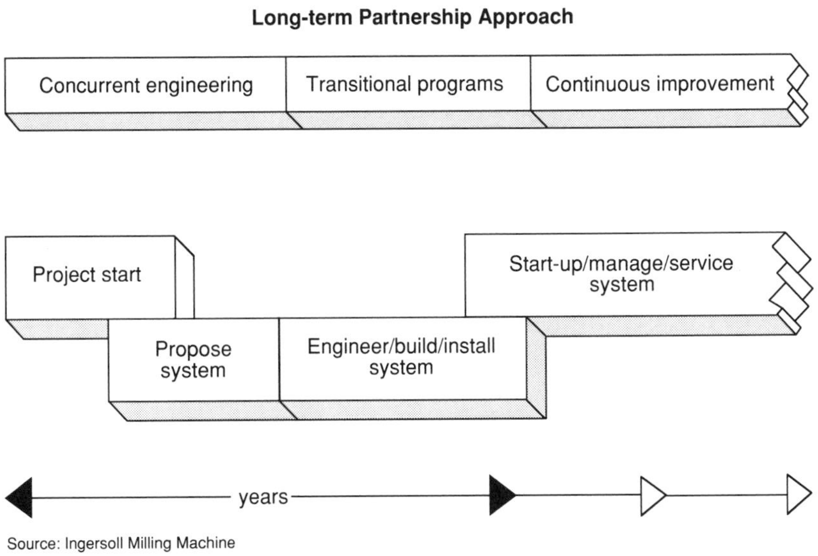

Figure 13-2. With a long-term partnership between vendor and manufacturer, improvements can continue for as long as the machines are in operation.

why relationships between automakers and the machine tool manufacturers are still subject to friction from time to time.

Reduced Lead Time for Plant

Nevertheless, with the right degree of trust, both the customer and the vendor gain from CE. The vendors gain more time to carry out simulations and test-runs of novel solutions, so their risk is reduced and they can develop new techniques. Because they have the information early, they are able to cut the lead time for the manufacture of the equipment.

The vendor of a transfer line now needs to give far more attention to running costs than in the past. It should not merely design the machine to meet the cycle time at lowest capital cost, but consider the labor involved in maintenance and tooling. As part of the team, the vendor is concerned with the whole life costs of production.

With CE, the product manufacturer saves in many ways. First, it gains much more commitment from the vendor than is possible if the design is completed and then bids are sought. Secondly, it saves time, since it no longer needs to go through the process of making inquiries to find vendors and then of evaluating competitive bids halfway through the project. That saves about three to six months, according to the machines and the way the project is handled. Further savings come because when the vendor starts detailed design of the line, it has much more information than would be available with the conventional method. Midwest Automation Systems (Buffalo Grove, Illinois) reckons that changes occupy 45 percent of all time involved in the design of automation equipment, so there is significant opportunity to cut costs here. In addition, when the vendor is involved in the CE team, it has found that six to eight weeks are saved because the vendor is not involved in bidding, and that four to six months can be shaved off the delivery lead time on account of the close liaison between customer and vendor.

In a conventional project to build a new engine, the total time from concept to the start-up of production is approximately 44 months, and the lead time is dictated by the design and manufacture of the transfer lines. A typical timetable would run as follows:

- Month 1 to 8: concept design by manufacturer
- Month 8 to 10: issue of quotations
- Month 10 to 13: vendors submit quotations
- Month 13 to 16: evaluation of quotations
- Month 16: issue of purchase order to one vendor
- Month 16 to 32: vendor designs and manufactures machines
- Month 32 to 36: installation
- Month 36 to 44: build up to volume production

Significantly, the vendor is not involved until the 8th month, and does not start to work on the machinery details until the 16th month.

With CE, the timetable is not only a little shorter, but starts earlier:

- Month 1 to 2: several vendors receive broad outline of requirements, and quote against a reference budget
- Month 4: letter of intent issued to one vendor
- Month 8: design optimized, and issue of purchase order
- Month 8 to 22: vendor designs and manufactures machinery
- Month 22 to 24: installation
- Month 24 to 30: commissioning, build up to volume production

In this case, the overall reduction in lead time is 14 months, although the actual cycle for design and manufacture is cut by only 2 months. The main reduction in lead time results from the earlier involvement of the vendor in the project. In a case where over-the-fence engineering led to many changes being made late in the program, the cut in the lead time could be 20 to 30 months, and is unlikely to be less than 12 months.

Apart from this tangible benefit, manufacturing also gains the expertise of the vendor throughout the project, and not just when quotations are being prepared. The close involvement mirrors that from the manufacturing engineers, with conceptual knowledge helping the machine tool makers improve the design

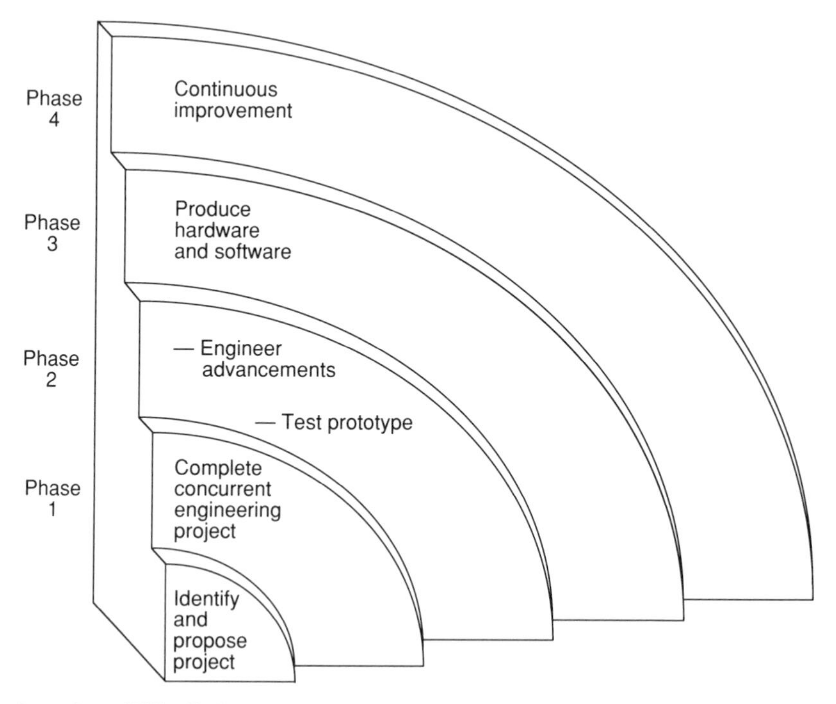

Figure 13-3. A machine tool supplier can be involved in four phases of a CE program, from concept to continuous improvement, after the product has been in production for several years.

without being told that design is not their concern, as often happened with over-the-fence engineering.

Some vendors of plant have complained that in the initial selection phase they are obliged to include details of their latest ideas, and that if they lose the contract, then the customer and their competitors gain details of their research. This is a real problem, and one that can be solved only with trust on both sides, and with the manufacturer limiting initial inquiries to companies it considers to be real competitors for the job. Adding three extra companies just to impress colleagues will prove counterproductive in the long run. In due course, manufacturers dedicated to CE will form natural long-term relationships with certain vendors.

More Information for the Vendor

Once involved in an CE project, the vendor of machines or a transfer line is continually asking for additional information and thus plays an important part in increasing the flow of information to the task force. If the vendor is outside the task force, its need for information does not have the same weight, and is often ignored until the design is complete. By that time, the vendor has a contract and is wary of upsetting the purchaser with requests for changes. Once in the team, the vendor can make any requests, so long as the customer will gain.

To demonstrate the advantages of CE in which the machine tool vendor is involved in the team, Ingersoll Milling Machine cites a conventional project in 1978 for a transfer line for a four-cylinder engine and a line for a V-6 engine in 1988. In 1978, the customer made 62 changes that added $1.3 million to the cost; with CE, the team made only 7 changes to the V-6 block, and they cost $436,000. Moreover, Ingersoll obtained changes to the specification through CE that saved another $750,000. On top of that, the project took 23 weeks less to complete than the earlier one, so that total savings came to a staggering $3 million.

In another program, Ingersoll Milling Machine was working with one of the U.S. Big Three in a program to develop a new series of V-6 engines and was able to work with the company to reduce the number of different sizes of bolts from 50 to 30 — some specified by vendors of components such as oil filters — and to change the tolerances.* The result was a reduction in lead time of 25 weeks, and a saving in capital equipment costs of $700,000.

Comau Productivity Systems reports a similar situation in which 13 sizes of bolts required to fasten the power train and suspension units to the body were cut to 2. Lamb Technicon quotes cases where standardization of the depths and diameters of tapped holes has cut costs significantly. It points out that as a specialist it has built up a substantial data base of the tolerances that can be achieved with various techniques, and so can advise the customer on the best approach. Also, the vendor is more up to date on the latest tools and techniques than the customer, and may be able to help attain tight tolerances with inexpensive techniques, simply because one of its staff is present at the meeting where the problem is first discussed.

Ingersoll Milling Machine, which has been involved in more than 20 CE projects with the automobile industry, is able to use concept drawings to provide outline costs. The concept drawings are continually updated, and usually Ingersoll is able to advise on the optimum positions for locating lugs well before the tooling for the casting is made.

It highlights several advantages gained as a result of its involvement in CE teams. In the case of an in-line four-cylinder block line, it was able to recommend far faster feeds and speeds

---

* D. de Lorge, "Role of the Equipment Supplier in Simultaneous Engineering," in *Proceedings of the 1st International Conference on Simultaneous Engineering*, pp. 137-155.

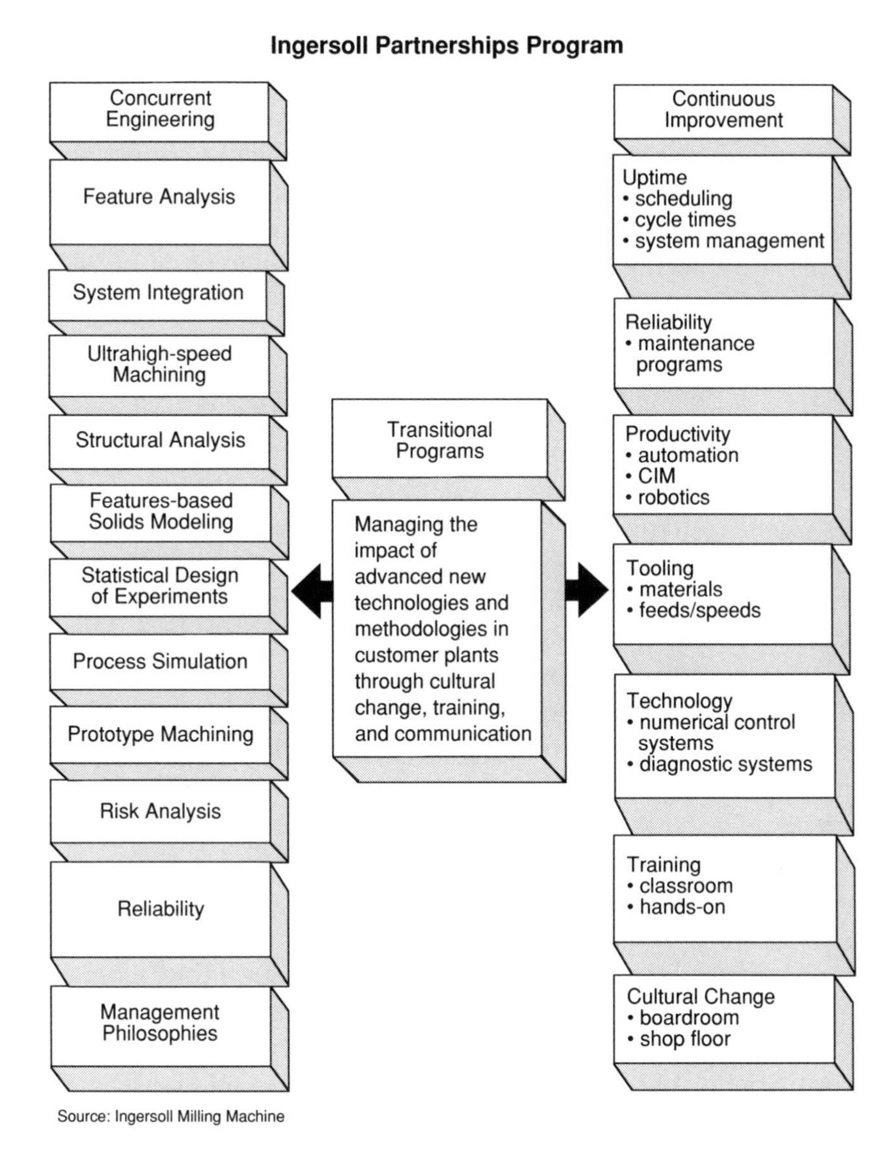

Source: Ingersoll Milling Machine

Figure 13-4. Ingersoll Milling Machine can provide the manufacturer with a range of services from the CE concept stage onward.

than the customer was planning, with a savings of $5 million — a cost reduction of 15 percent — and it cut two months from the program time. It also produced an unusual machine layout for a transaxle housing based on seven CNC machining centers and an overhead gantry to transfer workpieces between machines. It cost $1.45 million less than planned, mainly because fewer stations were needed, while an operating efficiency of 75 percent was achieved, against the 50 percent achieved by the customer's existing plant.

Substantial savings were also made with a V-8 cylinder block line as a result of improvements in machining techniques and changes in specifications and processes. The savings totaled $6.5 million.

To ensure that the vendor is involved in the line after the machines are installed, GM Engine Division employs one Ingersoll Milling Machine employee on four of its transfer lines. Paid by GM at an hourly rate, the Ingersoll employee's job is to make any improvements that can be made simply. He or she also reports back to Ingersoll regularly on what else needs to be done to improve operating efficiency. This is an ongoing program, without any contractual time frame, although the agreement can be terminated at 30 days' notice.

Early Simulation

Ford adopted CE and involved the vendors when it converted the Romeo plant to produce the 4.6 V-8 liter engine. Comau Productivity Systems was involved at an early stage in the planning of the transfer line, and was able to carry out a number of simulations, and assess the optimum arrangement of sensors and other equipment to reduce whole life costs. The contract for the assembly line was awarded to Wilson Automation (Warren, Michigan) whose chief engineer spent a year attending weekly meetings with Ford engineers to discuss every component with the team. The result of the involvement of both suppliers was reduced lead time and cost, with improved operation once the plant was installed.

Another example of CE can be found in the long-term relationship between Navistar International Transportation and Lamb Technicon. Navistar is using multidisciplinary teams to define product design and manufacturing processes, initially on a new diesel engine.

Navistar has passed through a decade of financial difficulties, and now consists only of truck and engine divisions. Its New Generation Diesel (NGD), designed to meet forthcoming U.S. emissions regulations, was the first CE project in the corporation and is due to go into production in 1992.

Shortly after the concept stage, Navistar asked a machine tool manufacturer to work jointly with it to review the design, to help with costings and definition of tolerances, and to lay out the transfer lines for the new engine. Although some improvements were made, Navistar realized it was not the ideal approach. In December 1987 it therefore signed a letter of intent with Lamb Technicon to collaborate over a 10-year period. The first project concerned four machining lines for the NGD engine, with installation starting in 1990. Uptime on transfer lines in the United States is 60 to 70 percent, but by using CE and working with Lamb Technicon, Navistar is aiming for 85 percent.

Automation Became Feasible

Another example of the benefits of working with the vendor is demonstrated by the way in which John Brown Automation worked with Ford of Europe from 1987 to 1989 on a new pedal and pedalbox assembly. Initially, one engineer from John Brown Automation attended the task force meetings, which were held every three or four weeks, while the company carried on its development work in its own factory. Toward the end of the project, an engineer from John Brown Automation worked in the Ford design office as prototypes were modified to improve manufacture.

The result was a set of components that could be assembled automatically, whereas the previous components could not. The number of parts was reduced from 13 to 8, less machining was involved, and the total manufacturing time was reduced from 6 minutes and 20 seconds to 4 minutes and 34 seconds.

To increase trust, and to gain the most from CE, vendors need to change their method of quoting for jobs, and to modernize their approach to maintenance. The method adopted by Comau Productivity Systems overcomes many of the problems. Its prices are based on the building blocks of the transfer line — irrespective of whether this consists of special-purpose machines or CNC machines and ancillaries. Its quotation is therefore based on the use of a number of modules adding up to a certain price, with a certain sum being allocated for services such as swarf removal. If the design is changed so that one less module of one type is required, but three more of another, the customer will be able to see immediately how the price will change. This approach, which builds trust between manufacturer and vendor, is an asset in CE projects.

Attention to Maintenance

Another area where changes are required is in maintenance. Many manufacturers complain that the currently popular concept of planned maintenance has proved too costly. They have to stop the line at regular intervals to replace specified components, often finding that the components are quite serviceable; so the efficiency is reduced by this extra downtime, and they have installed new components at extra cost. Another approach is to load up the transfer line with sensors that indicate when components need replacing. However, these sensors, which increase the cost of the line quite substantially, can fail, increasing the downtime further.

To overcome this problem, it is not enough to predict the life of components, such as bearings, that are subject to wear. It is preferable to supply the customer with diagnostic data and instruments to measure the machine to determine when renewal is necessary. For example, the vendor should be able to supply an instrument to measure the vibration levels of bearings, with charts to show acceptable, dangerous, and catastrophic levels. The customer can then minimize downtime and maintenance costs.

Better Component Design

Vendors of critical components, particularly specialists such as foundries, electronics manufacturers, or suppliers of fuel systems, are included in the task force and encouraged to consider whole life costs. As mentioned in Chapter 4, GM now involves its vendors of steel sheet in its task forces, and has therefore been able to improve the "friendliness" of the manufacturing process. Clearly, vendors of sheet steel and castings are key suppliers as well as the makers of specialist components. Early involvement in the task force benefits both manufacturer and vendor.

As mentioned in the previous chapter, because of the complexity of the forms involved, the foundry responsible for the casting of a cylinder head or block or transmission casting should always be involved in the task force. In the case of a cylinder head, the engine designer works hard to optimize a number of dimensions, forms, and parameters such as the combustion chamber shape and port shape. Although the designer is not really concerned about the detailed shapes in some areas, he or she usually designs every detail without consulting a foundry specialist until finished.

Cylinder Head Lead Time Cut by 60 Percent

For this reason, one of the large automakers set up a team involving Alcan to design the aspects of a new four-valve head that affected the foundry.* The engine designer supplied the foundry with the CAD data for the main dimensions and forms. He specified the basic dimensions of the head, and the relevant dimensions of the cylinder block, the retaining bolt positions, and data including

- location of spark plugs
- valve train geometry
- shape of the combustion chamber
- shape of the ports
- arrangement of coolant passages and position of thermostat

At this stage, of course, although some development of port shapes and compression ratio could continue, the basic form of the head was fixed; the fact that it was for an existing engine determined the bolt fixings and general layout of the bottom face anyway. The vendor of the cylinder head was responsible for ensuring that the head could be cast with a high yield and was to design all details related to the foundry, such as runners and risers. The same vendor was to optimize the cores, which involved responsibility for the internal design of the head. It also undertook to ensure that engine design parameters such as cooling and ease of machining would be met.

All this was done on a concurrent engineering basis, with frequent meetings of the team and exchange of CAD data. As soon as the foundry experts started work, they found that the change

* B. Eigenfield, B. Kroeger, and T. Ortleb, *3D CAD/CAM Design of a 4-valve Cylinder Aluminum Head,* SAE Technical Paper 900655 (1990).

in section, and mass of metal where the inlet ports met the bosses around the spark plugs were unacceptable. Therefore, even before they started their work, they had to ask the designer to revise this arrangement. Wall thicknesses were agreed on at this stage by designer and foundry, with a general wall thickness of four millimeters, but with a nine-millimeter thick bottom plate, and numerous intermediate thicknesses in specified areas.

The second problem encountered was that the exhaust flange to which the manifold was to be attached was shaped so that it would restrict coolant around the exhaust port. This area was redesigned, and there was a considerable amount of joint work in the design of the coolant passages. Normally, the problems would not have been found early enough in the project for this to have been done.

Design and Prototype in Three Months

Because the customer and vendor were using common data for their work, the whole process was much quicker than normal, with design completed and prototypes supplied within 3 months. Normally, 8 to 12 months are needed. Among the reasons for the speed was the fact that as soon as the port shapes had been finalized, the CAD data could be used to drive CNC milling machines to produce the masters for the patterns. Likewise, with the combustion chamber, bolt positions, and coolant transfer holes known, the bottom plate for the molds — in steel — could be made even though the internal form of the head had not been completed.

One of the most difficult areas to design — on CAD equipment or manually — is the water jacket. With CAD, it proved practical to design a basic form for one cylinder and then modify it to produce the form at the front and rear of the head. This part of the design took five weeks, whereas the tappet chamber took three weeks.

Because the foundry specialist was responsible for much of the design, he worked on it in the sequence that the foundry would need to produce the prototype patterns. Thus, the bottom face was fixed early on, the ports were designed as a module, the outer faces — hence the need to fix the exhaust flange early — as another one, and then the tappet chamber in the top of the head. CAD played an important part in the reduction in lead time from about nine to three months, but the team effort was more important in allowing the production of the cores to proceed before the design was finalized. This approach should be adopted for all similar components, whether the foundry is in-house or a vendor.

The same benefits are obtained when the components are produced completely by a vendor — whether those be shock absorbers, instruments, electronic controllers, microprocessors, computer motherboards, or cables for cranes. In one case, Lucas Automotive, which operates CE for its own projects, was brought into a team at a customer's product development department and asked to develop a component for a high-level model. Volume was to be low, at a few thousand a year. Initially, when the team looked at the product they thought it was impossible to produce it at such low volumes for the required price.

QFD Ranks Features

In the old days, that would have been that; the team would have told the customer's purchasing officer, and the chances are that the product would have died. But when the members of the customer's task force were told of the high price, they looked at their QFD matrix and found that price was low on the list of the customer's priorities. They went back and checked with potential customers again, and decided that the price was acceptable.

Meanwhile, the marketing person on the Lucas task force realized that with some minor modifications and cosmetic changes it might be possible to sell the product to other manufacturers. This proved to be the case, so that well before the design was frozen, potential demand was up to several thousand a month, with the result that the price would fall well below the initial target! In cases like that, CE compounds its gains.

Once a vendor of a component is involved fully in a project, the manufacturer can abandon some bad practices, such as forcing a detailed design on the vendor. Naturally, the normal process is for the customer to send a drawing to the vendor. With specialists, sometimes only the outline is shown, but more often than not, the vendor's drawing is copied onto the customer's standard drawing sheets. When a succeeding product is required, the customer then modifies the old drawing and sends it to the vendor.

Functional Design Specification

However, many details transferred from the old drawing may be inappropriate. For example, if a windshield wiper arm is merely scaled up, it may be out of proportion. Similarly, if an exhaust system is required to mate up with a tubular steel exhaust manifold instead of a cast iron manifold, it will need to be redesigned completely. For these reasons, manufacturers should specify only the design requirements, in terms of function and performance, as well as the restraints on size. This approach, which will release new ideas and processes from the vendor, mirrors the basic design concept of CE in which the product is specified in the customers' voice. The customers will need to test the product exhaustively — preferably in conjunction with the vendor and, of course, under the control of the task force.

CE liberates both manufacturer and vendor from the adversarial attitude fostered by purchasing. Instead, the two companies can work together on a new design with the same overall targets. The targets are

- optimum design
- optimum quality
- lowest piece cost and capital costs
- shortest time to market

Every vendor will gain from the success of the product. Machine tool builders will be in a better position to gain repeat orders or orders for similar equipment from other manufacturers, and will gain much more information on the actual performance of their equipment than in the past. Component vendors will gain the increased volume that comes with success, and through their experience are likely to gain follow-on orders.

Some personnel in manufacturing will be suspicious of any approach that seems to give the vendor carte blanche to supply a product. CE does not attempt to do that. However, involvement in the task force will give vendors an advantage over their competitors, so their product and quality should be superior. If they are not competitive, they will not last long.

Successful vendors will be achieving world-class quality, and manufacturers will not want to jeopardize that by replacing an established vendor with another that promises a lower price. Indeed, in an CE environment, the purchasing manager will not be permitted to change a vendor without evidence that the new company can actually supply quality products in volume for a lower price than the existing one, so each vendor is safeguarded; so long as its quality and prices remain competitive, it will get the order.

Therefore, CE does alter the relationship between manufacturer and vendor. Furthermore, it takes the current trend of moving from multiple to single sourcing into a new world full of opportunities, and lays to rest the adversarial approach once and for all.

PART THREE

Making It Happen

CHAPTER FOURTEEN

Starting Concurrent Engineering

Concentrate continuously on changing culture
Set targets
Train staff in methodologies
Start with small project and build up

At first sight, the changes needed to introduce concurrent engineering may seem daunting, and suitable only for very large companies. Fortunately, CE lends itself to a gradual introduction, and is as useful on small projects as on major ones. Thus, small companies as well as large ones can pick up the elements they need. Indeed, the smallest companies are likely to be using some elements of CE simply because employees often must juggle more than one responsibility.

In implementing a CE program, the first requirement is that the lead be taken at board level. The program should not be left under the control of the vice president of product design; once the plan is accepted, the chief executive should make it clear that he or she is the ultimate leader. However, the chief executive's role is to push the plan forward and to ensure that it does not become the preserve of one department, such as product engineering, even though it is normal for a product engineer to lead each task force.

In some cases, it may be sufficient to instruct product design and manufacturing engineering that they must rethink how they develop products. In others, it may be necessary to change the structure of the company. One approach is to define strategic business units, as Delco Remy did, and then to assign product designers, manufacturing engineers, and other specialists to each.

Alternatively, it may be best to make one vice president responsible for both product design and production engineering, or to restructure product engineering so that product designers and manufacturing engineers responsible for one group of products — such as power units in the automobile industry or semiconductors in an electronics company — are in one division. Restructuring makes it easier for people to accept other changes.

Make Bold Moves

Bold moves are needed to drive home the need for a new way of thinking and acting. Changing the way people think about product development is the key, and must be brought home to people at all times. It has been mentioned several times in this book that the culture of the corporation needs to be changed, and most exponents admit that changing corporate culture is the most difficult part of implementing CE. Some say that CE itself is a simple concept, but making it work is difficult.

Certainly, the Big Three automakers have spent a lot of time using CE without really changing the corporate structure. Ford has been attempting to weld CE into its structure and into some other approaches, such as its "concept to customer" program for some time.

One thing is clear, however: if serious efforts are not made to change the culture, especially in a large corporation with entrenched customs and organization, CE task forces are likely to end up as weak committees, still under the thumbs of the department heads.

There are two aspects to this culture change:

- Changing the attitude of all personnel toward quality and business priorities
- Changing the culture so that sufficient power is vested in the task forces, which means taking power away from functional departments

Personnel in the corporation must be made aware of the difference between what *they* perceive as quality and what *the customer* perceives. They need to take note of their attitudes when having their car serviced, or buying some shoes, furniture, or other products. For example, when a project engineer, acting as a private individual, wants to buy a dining table, he or she looks for a durable piece of equipment, of a convenient shape and size, and with some style. Similar considerations are made when buying a car. When the engineer receives the table ordered, the first thing he or she does is to inspect the finish — not the shape, size, or durability.

If there happens to be a small scratch near the bottom of one leg, the table will go back, even though the durability, convenience, and style of the table remain unchanged. But the engineer does not think that there should be a scratch on a new piece of furniture, even though Little Billy is likely to crash his model truck into it the first chance he gets. This attitude is an example of the difference between quality as seen by the customer and the seller.

Taking the Customer's Viewpoint

Let's say a man who works as a project engineer goes to collect his car after it has been serviced. He sees that in addition to the $254 for the parts and labor, there is a $10 charge for "sundries." When questioned about this, the service reception manager loses his cheerfulness and says that it covers things that cannot be itemized, like cleaning rags and grease. The customer's complaints are to no avail, and he goes home feeling aggrieved. Yet the same man will wonder why a major OEM customer sent back a component with a discoloration on an area of the product that was normally hidden.

There are two sides to every story of poor quality, because the manufacturer or supplier either does not know what the customer wants or considers that it knows best what is acceptable. But if a customer complains, there is a reason for doing so, and next time he or she is likely to go elsewhere for the product or service. It is important to recognize that customers cannot reassure themselves that they have bought high-quality products by running a series of durability tests. The only thing they can do is to check the superficial quality. For that reason, finish is far more important than it seems. With products shipped out in cartons, quality begins with the appearance of the carton, and the way the product is packed in it. Is it secure? Can the product be removed easily? Are all the components listed, and easy to see? This is not just a question of public relations; as customers open the box they are gaining their first impression of the quality of the supplier.

Nor are superficial changes such as persuading the receptionist to say "How may I help you?" and "You're welcome" to every customer much help. Of course, politeness is necessary, but the actions taken to solve the customer's problem are more important. What does this have to do with CE? CE is about putting the customers and their requirements first, and providing real service throughout the life of the product. It often involves a major cultural change, since it must pervade the corporation from top to bottom.

Program for Change

To foster this new culture, a program for change — in which the importance of quality is highlighted — is needed at all levels. But shouldn't this have been done when TQC was adopted? A start may have been made, but re-emphasis with the accent on understanding the customer's viewpoint will pay dividends. In any case, many companies have not implemented TQC, or have not done so with total commitment.

As a prelude to its involvement in CE, Digital Equipment, where the program was pushed through by an enthusiastic vice president, disbanded its quality control department. This was a symbolic gesture and an important prelude to restructuring. Once that step had been taken, and personnel had realized the implications, they were prepared for change. With quality now considered the responsibility of all employees, the people formerly in quality control have been deployed in different departments where their experience is of greater value. As mentioned earlier, a change in the grouping of product design and manufacturing can also help.

Power to the Task Force

Once a corporation has established a culture for change with the accent on quality and business performance, it will have a base on which to build CE. The chief executive, or one of the senior vice presidents, should make clear his or her total commitment. In a small or medium-sized company, this person may head the CE steering committee as well. The steering committee will be responsible for deciding how CE is implemented, and thereafter for modifying the system in the light of experience. It needs to be led by an CE champion — someone who is dedicated to the concept and prepared to stake his or her job on its success.

At this stage it is important that an infrastructure be developed to support CE. It is inevitable that as the changes are being made, CE will hit a few snags, ruffle a few feathers, and create some opposition. Some managers will grumble that it does not work, because they want to go back to their comfortable routine and not lose power to the task force. At that point, the problems need to be solved quickly, and be shown to have been solved. At the same time, personnel should be left in no doubt that the development of the new system will lead to early problems, but that these will be solved.

If the company is fortunate enough to avoid such internal criticism, the steering committee must still continue to work to increase the scope of CE. It is also important that staff understand why CE is being adopted and that it will evolve continuously, and will not be dropped in favor of a new buzzword in a few months.

The next change is to set up a pilot project and give the task force sufficient power to pursue it successfully. It is helpful if the task force is carrying out its work on a site remote from the mainstream activities, so that team spirit can be fostered and members are unlikely to be dragged away to help on other projects. This was the approach adopted at Delco Remy, as mentioned in Chapter 4.

It needs to be understood that in pushing through the task force concept, the management structure is of no help at all. Therefore, big corporations that pride themselves on an efficient structure around which everything revolves find this change difficult. But it is a change that must be made.

In parallel with the setting up of the first task force, the steering committee should establish a training program for all personnel likely to be involved in CE. Initially, the training related to working in a task force will be concerned with the philosophy — and the changes in culture. As time passes, examples of best practice will be written up and converted to training modules for various aspects of the process. Eventually, modules covering every aspect of the process, from preconcept to product launch, can be produced — a process adopted by Xerox. These will not be step-by-step guides, but they should lay down the principles and the essential stages.

Training is Vital

From the outset, the company must train people to use the essential features of CE — two days in QFD, DFMA, and Taguchi is a good start. Then, product design engineers need to be trained in production processes; the amount of detail will depend on the size of the company. At the same time, people should be encouraged to take a practical approach to their new roles by looking at the products and production processes. They need to understand the product in much more detail than they think necessary, and need to take a broader view of their responsibilities.

Once in the task force, they will have a wider role to play than that of a product designer or manufacturing engineer operating in his or her own department. Because product design engineers often head teams, and because many designers feel threatened by the presence of people of other disciplines in the team, training should be concentrated on design engineers.

For example, when it decided to adopt CE, Smiths Industries Aerospace & Defense Systems, Inc. spent the first year in training its personnel. DFMA, design for cost competitiveness, cross-functional teams, and the need for reducing lead time were included in the courses.

Manufacturing engineers need to be trained to improve the efficiency of their equipment with the application of DFMA and FMEA, and to apply the same whole life approach to their plant as the task force applies to its product. Everyone in the task force and back in the functional departments must now be concerned about waste of time and any operations of equipment that do not add value. Only by adding value in each operation can the company generate the profits it requires to improve the product and keep abreast of the competition. In this context, adding value has a broad meaning. A turn-over fixture does not add value itself, but may do so if its use allows the operator to do some work more efficiently. Likewise, time spent listening to the problems of the vendor may add value if the result is that each party gains an understanding of what the other is trying to achieve.

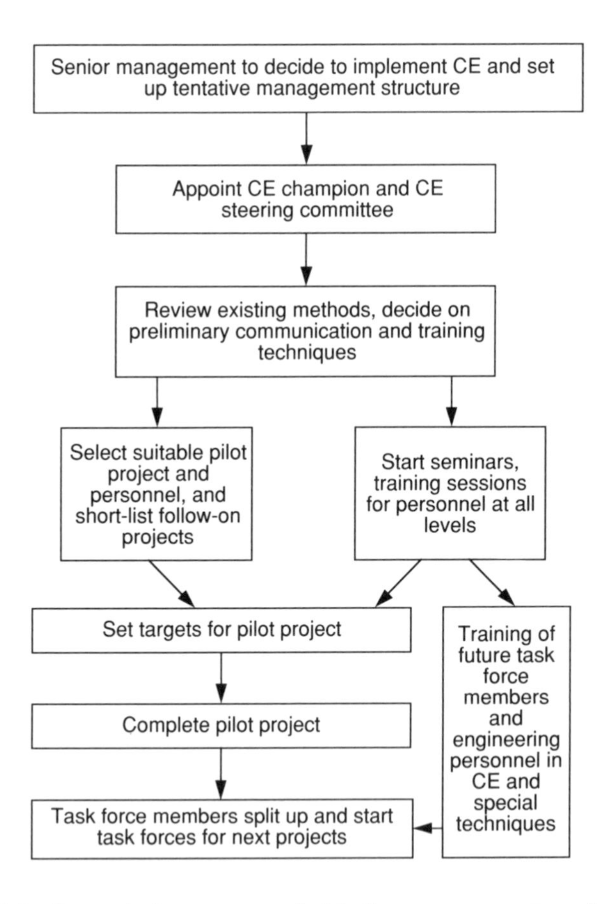

Figure 14-1. Several steps are needed before concurrent engineering is implemented, and the process starts with the total commitment of senior management.

Therefore, manufacturing engineers and managers should question aggressively every item of equipment that does not add value — such as forklift trucks, conveyors, buffers and stores, and equipment such as turn-over devices. They should always ask whether by changing the design or making some other modification, these can be eliminated, reduced, or simplified.

Manufacturing engineers must involve the vendors of the machine tools in the task force, and either build or commission the building of test cells to evaluate new methods of manufacture. The idea should be to build a rough-and-ready cell quickly, rather than an elaborate semiproduction unit. The engineers should do the same when they have doubts about whether a device in their own system is really needed.

Targets Essential

Targets should be set from the outset; these may be unattainable, but they will concentrate the minds of task force members on what they are trying to do. For example, when it moved into CE, Chrysler established a target of reducing the total unit cost of each vehicle by $2,500. Lucas Automotive established the following targets:

- 50 percent reduction in lead time
- 50 percent reduction in costs
- Zero defects
- Simplified control procedures
- Design standardization

These are typical targets, and cover important areas. Each company will need to consider its own priorities, but whatever they are, an overall upgrading of efficiency in manufacture and documentation of changes is a necessity.

Three Types of Task Force

There are different views on whether there should be a true leader of the task force. Some say that any attempt to impose a manager on the team is counterproductive, and that all members should be equal. One executive in a large U.S. company with considerable experience in CE said that there are no memos, and no decisions are postponed because of disagreement — of course, they may be postponed for lack of information or because there is plenty of time and more pressing matters. However, in that company, if two members of the team cannot agree, they are left to argue it out and come to a decision. Needless to say, this is where the others do not think they can add anything to the argument. This is a difficult approach for large companies, and for Americans in general, to adopt. They are used to the concept of strong leaders, and the success of individuals, not the success of a group. One member of the team may be appointed leader, but the leader is more like a chairperson, whose job is to give the meetings some structure and to prevent time being wasted. He or she is not a manager.

Others take a different view and look for potential leaders for their task forces. They want a strong character who is dedicated to the concept, but not a domineering individual. This type of leader must be able to draw the best out of people, each of whom has with different characteristics. Above all, he or she must be a team builder. If the quieter members of the team believe that their ideas are never given a chance, they will soon cease to be useful members.

Leaders must first and foremost be dedicated totally to the concept of CE. Second, they must be skilled diplomats. They must be able to turn half-baked ideas into solutions by ensuring that everyone takes the ideas seriously. If the designers and manufacturers cannot see things eye to eye — not so uncommon in the early stages with CE — the team leaders need to draw them together and show that if they all consider an idea — however impractical it may seem at first — in a humble and professional manner, an improved design may emerge.

When the designers serve as team leaders, as often happens, they must remain receptive to ideas from outside the design department. If they see design as the special preserve of their department, then the team will not be a real team, and they should not be in the team, let alone serve as leader. Interestingly, Delco Remy appointed a product designer as leader of its first CE team, but a manufacturing engineer as leader of the second team, which was set up at almost the same time — a sound, evenhanded approach. In companies that must continuously respond to customers' changing whims, it may be appropriate for a marketing manager to lead the team.

Clearly, task force leaders should not seek to dominate others, but must establish an environment in which ideas will be nurtured. They must also keep people moving ahead to maintain the schedule.

First Project

Once the company has decided to adopt CE and create a task force to handle a pilot project, the next step is to decide on the type of task force. There are three basic options:

- A preconcept team of four or five people — from product design, manufacture, marketing, and probably finance and service — which is expanded to a full task force once the concept stage is reached.
- A full task force to take the product from preconcept to production. It is important to establish as the cut-off point satisfactory full production, not Job One, so that the team is aware of the problems in the early stages of production.

- A task force that starts at preconcept stage and then continues as long as the product is in production. In this case, the task force is likely to be a profit center or business team, accountable for the overall profit of the product from start to finish. The size of the task force will vary according to the stage of the project. It will start small, grow to full size after the concept, and then slim down again once production is reached.

When CE is new, the first approach is probably easiest to handle, and the actual number of people involved will depend on the complexity of the product. Ten to 30 people is a realistic number to work together, but in the case of a completely new automobile, there are likely to be more than 100 people, divided into teams for engine, transmission, chassis, body-in-white, body trim, and electronics/instruments.

Clearly, it would not be wise to start a CE program by adopting it for a completely new vehicle or family of computers or dumpers. A new component, a new model derivative, or a face-lift for an existing model would be more appropriate. In consumer electronics, an upgraded model might be appropriate, and in computers a peripheral product such as a disk drive or a set of housings are suitable. In all cases, the aim is to limit the project to a manageable size until the problems are understood.

Too much should not be expected from a pilot project. Because it is really a training exercise, it is a good idea to build up the team from the nucleus to the full size quite early on, and to alter the membership once the product approaches manufacture. Thus, as many people as practical will gain some experience in CE, and will be able to play a fuller part when more teams are started up. Nevertheless, the gains made by the pilot project (and subsequent projects) must be publicized within the company so that personnel will realize that CE works and will therefore want to be involved. In this way, the benefits will be compounded.

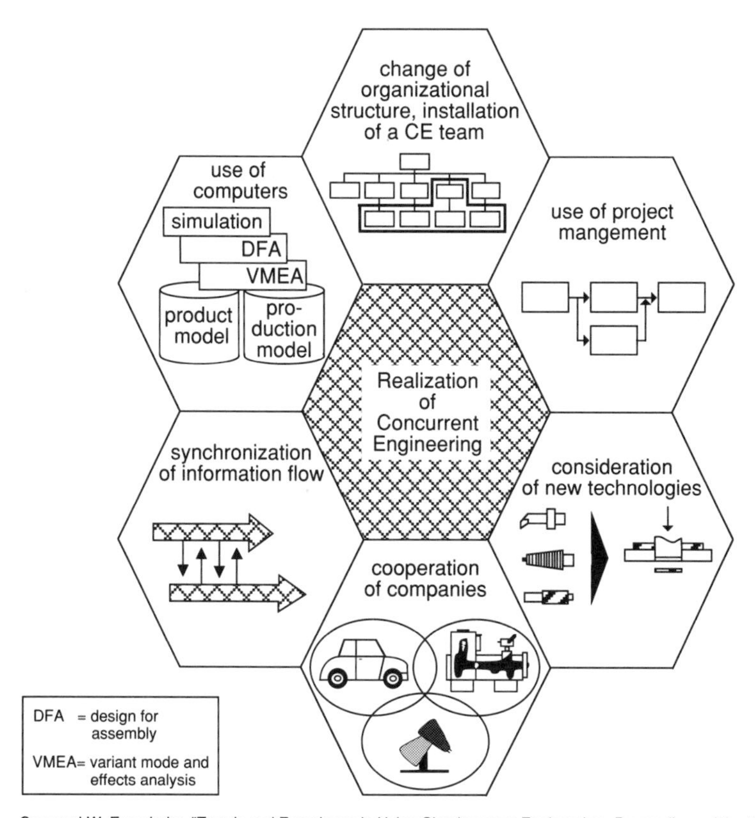

Source: I.W. Eversheim, "Trends and Experience in Using Simultaneous Engineering, *Proceedings of the 1st International Conference on Simultaneous Engineering*, London, December, 1990, p.18.

Figure 14-2. In setting up CE, it is necessary to coordinate the exploitation of new manufacturing techniques, management methods, and information technology.

Maximum benefits are obtained in long and complex projects, where there is more scope for misunderstanding between departments. Any major structure, vehicle, or machine is suitable, and there is no need for new technology or manufacturing techniques to be involved.

The ultimate aim is to build up a set of teams, one for each product line. This will take time, and to ensure that the skills required for working successfully in a team are fostered, additional people can be co-opted into the teams for short periods. All these movements should be overseen initially by the steering committee, which will need to cooperate closely with leaders of the task forces until all are working efficiently.

Nor should individual components be ignored. In one case, a company wished to alter the appearance of a car by reshaping the plastic bumpers. However, since the change was for a facelift, the company did not want to buy new molds. Therefore, a small team consisting of stylist, design engineer, and injection molding expert was set up to obtain the maximum visual difference with the minimum changes to the molds. They were highly successful, with the front end of the new car looking completely different from the previous model, yet the cost was low.

Meet Customers

It is also important that right from the start, the task force goes out to meet the customers, and find out what they want. The whole team should go — it is just as important for the production engineer and financial officer to do this as it is for the designer. Once they start to do this, they will become a team, and their attitude to the product will change drastically.

As a small starter project, a company that supplies many different companies should investigate the time taken for orders placed by the customer to find their way through sales to the production schedule. In most companies this is a major source of delay. It does not have much to do with engineering, but it makes the team think about problems that are outside their normal sphere of influence, and helps them work together.

Another approach is to start with a project where the sole aim is to add a derivative to take customers from a competitor. This

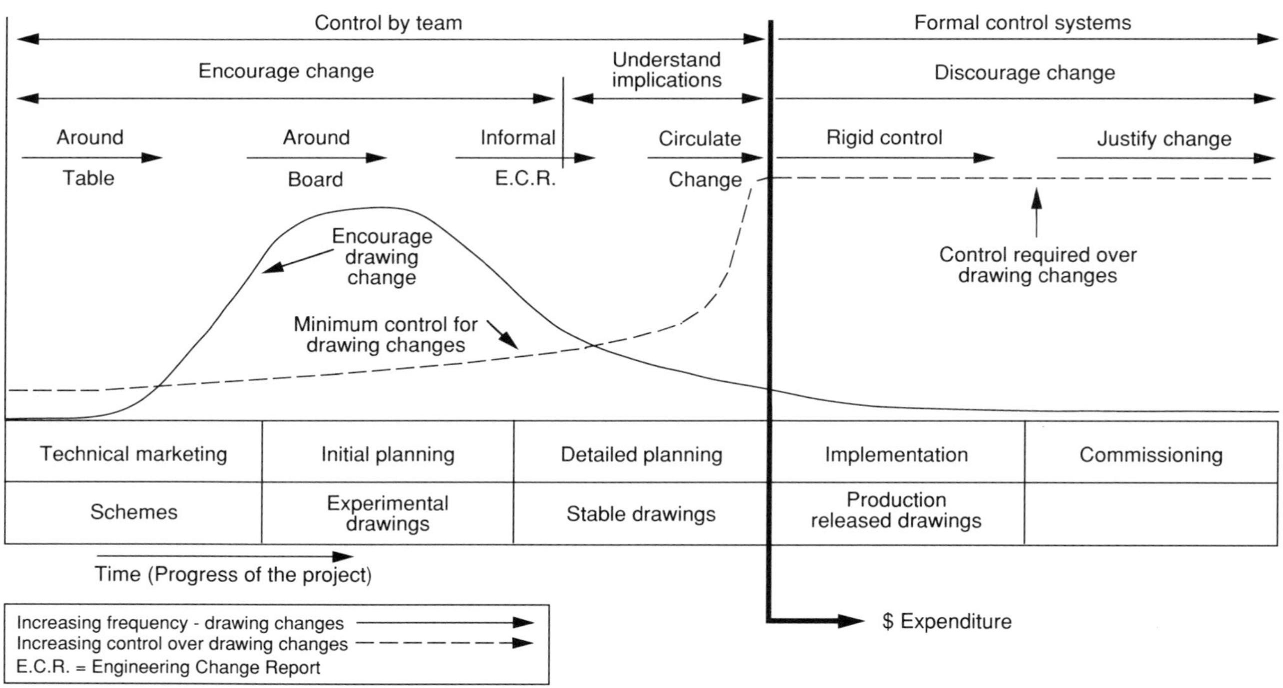

Source: J.C. Ford, "Simultaneous Engineering (Design to Manufacture)," in *Proceedings of Auto Tech,* Birmingham, England, November 1989.

Figure 14-3. At all times, members of the task force need to work to prevent changes being made late in the program.

immediately takes the team into a new environment. They must find the customers; they cannot rely on their dealerships or the opinions of sales as to who are their customers. They need to find the real customers, gain their views, and compare those views with their own preconceptions.

Regular Meetings

Although CAD/CAM and a data base are important aids in concurrent engineering, electronic data interchange (EDI) is not a substitute for meetings. It might be thought that instead of holding actual meetings between the members of the task force, they could just interchange data through a central data base. However, as mentioned elsewhere, that approach will turn the operations into over-the-fence engineering under a different name. The designers sitting at their terminals will be reluctant to send their data until they have finished; the manufacturing engineers will be equally reluctant to give suppliers of machine tools incomplete data, and so sequential engineering is inevitable. The members of the task force must be just that — members of a team working together. Of course, they can never work all together on all things — at best, two members of the team will be working together.

Indeed, one of management's main tasks in establishing an CE environment is to remove barriers to CE. A physical barrier, with the members of the team working in different sites, is a cardinal sin, and a barrier that must be removed— hence the talk of colocation of teams. Another barrier is excessive bureaucracy. Delco Remy found that elimination of physical barriers and the establishment of an environment of minimal bureaucracy were essential in fostering CE.

Small companies do not need to worry that they are too small to introduce CE. In fact, it is the largest companies that have the most problems, because they have well-established hierarchies and management structures that will need to be changed, a move that will arouse opposition from many managers. By contrast, in a smaller company, the less rigid management structure allows decisions to be made quickly, and the few number of teams can be managed more easily.

CHAPTER FIFTEEN

Managing Concurrent Engineering

Emphasize need for change at all times
Concentrate on attaining targets
Add training modules

Once a company has introduced CE, it needs to maintain the impetus of the new philosophy and set up a task force each time a new project is started. The management structure will need to be changed to accommodate the new method of working. Reorganization of offices or the erection of new buildings to house the task forces is also beneficial and signals that senior management is dedicated to CE on a long-term basis. That dedication is vital, as is its evidence to all personnel.

Targets are essential, and those set at the beginning need to be reviewed and updated regularly. For example, once the new model introduction cycle has been cut from say, 54 to 48 months, or from 24 to 18 months, a new target involving a reduction of a further 15 to 25 percent should be set.

Personnel need to appreciate that the aim of the "new" culture is continuous improvement, that the commitment to CE is real, and that it will now become a way of life. Those who criticize the technique will be encouraged to devise ways of improving it rather than suggest a return to the bad old days — which quite a number of people will actually remember as being cozy and undemanding.

Guidelines for Managing CE

Commitment to CE and to changing the old methods is the most important factor in managing the new approach. Next in importance is the provision of resources needed by the task force, and a backup in the form of a steering committee responsible for updating methods as necessary.

There are a number of important guidelines, some concerned with management, and some with methodology. Some of the important ones are listed below.

1. Senior management needs to take a hands-off approach to the task forces, but must make its support for the system crystal clear. Vice presidents should be ready to foster whatever changes are needed to improve the effectiveness of CE, and whatever investment in training and new equipment is required. Regular bulletins should be posted on boards in offices and plants so that people are aware that CE is ongoing.

   Management must emphasize that now every action in the corporation has but one goal — to improve business performance, which ultimately means greater profits. For example, investment in CAD should increase efficiency in the design office. It will not do so immediately, but users must be trained so that within 6 to 12 months, a definite improvement is noted. This applies to any investment. It is not enough for the plan to state that it will improve efficiency; it must be proved to do so.

   Management will also need to plan for the long term because once lead times are reduced, new product programs will tend to come around much more quickly than previously. For example, with a lead time of seven years, new models will be introduced at seven- to eight-year intervals, with upgrades every three to four years. With four major product lines, one new project and one minor project will start every two years. With a 36- to 40-month lead time, one major and one minor project will start every year. Personnel in all departments will need to alter their way of approaching new model projects to cope with this shorter timing, and management will need to ensure that they are able to do so.

2. Strong leaders who are dedicated to CE should be chosen for the task forces. They will usually be from product engineering, but some companies appoint the plant manager or one of the functional managers as leader. In any case, this person is a leader among equals. Members of the task force will all be achievers, or potential achievers.
3. A formal structure for the task forces is necessary so that members understand their job functions and can work together; they should not remain in their own departments, attending weekly or monthly meetings only, as if they were attending any other product meeting. Ideally, team members should work together in the same office or in such a way that the team spirit is fostered. Some users adopt the "60-foot rule" (Northrop recommends the "100-foot rule"), the number indicating the maximum distance that separates desks of members of the team from one another.
4. A document of intent should be drawn up indicating the targets set by the task force. Team members will have agreed to those targets and so will be committed to attaining them. Review dates will be included — usually quarterly — and between those reviews team members should be left to manage their project unhindered. They should understand that every action they take should be focused on improving profits and market share.
5. Task forces should hold regular full meetings. In larger companies, the task force will consist of several smaller task forces, many of which will be working side by side most of

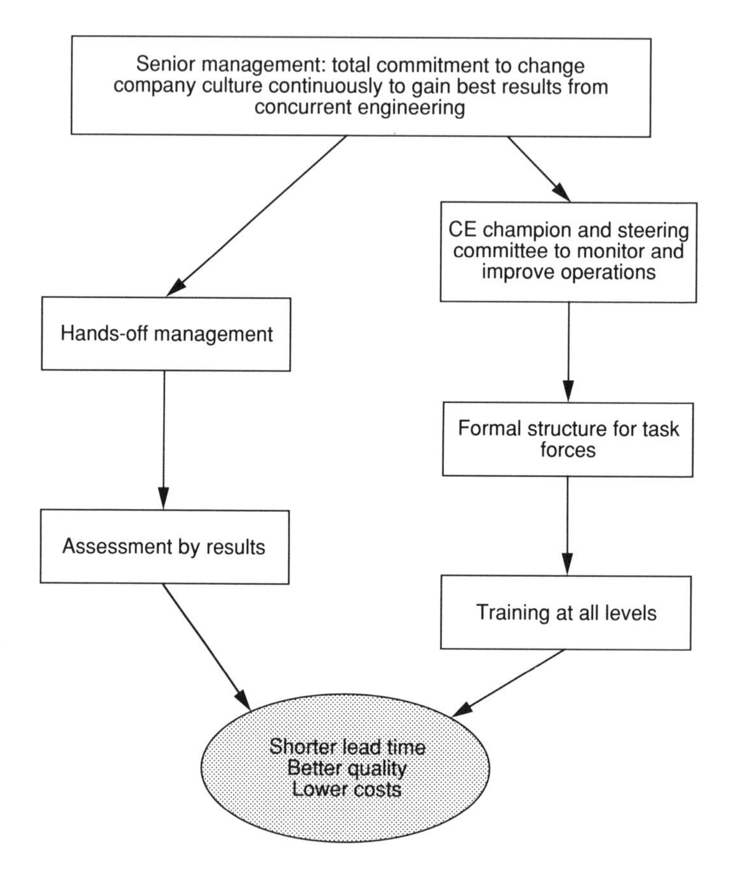

Figure 15-1. Essential stepping stones to the successful management of concurrent engineering.

the time. But it is important that the full task force meets frequently, so that all members have a grasp of the complete project and its progress.

6. The full task force should spend considerable time with customers. If this job is left to marketing or product design, the people involved tend to bias the results to match their own preconceptions. It is clearly not practical for 150 team mem-

bers to descend on one customer, but they can visit different customers or make the visits at different times.

Where a major project in the automobile or construction equipment industry is involved, some members could spend a day working in service reception at a dealership, or in the sales office. A manufacturer that has a close relationship with its dealerships has an opportunity for direct contact with customers.

7. To ensure that the members of the task force are fully conversant with the product or concept, the Honda techniques of *zenbara* (full dismantling) and *genba gembutsu* (actual place, actual product) are recommended. (See Chapter 3 for details.)
8. Training is an essential feature of CE. Methodology, controlled by a steering committee, should be established for the different techniques required and training sessions set up for

    - the task force approach
    - basic techniques, such as QFD, Taguchi, and DFMA
    - the need to get it right first time
    - what to measure to get it right
    - the various stages in the project

9. Product design engineers should also be trained in limitations and details of relevant production techniques. Manufacturing engineers should be trained to approach their work from a viewpoint of added value.
10. Purchasing managers should be retrained to take a wider view of the cost/performance relationship of products and machines. They need to understand more of the process in their plants, and at vendors' plants, and need to be able to form long-term relationships with vendors to maintain the mutual benefits sought at the outset. They should be steered away from the idea that cheapest is best, and that the vendor's representative is only seen when prices are to be discussed.

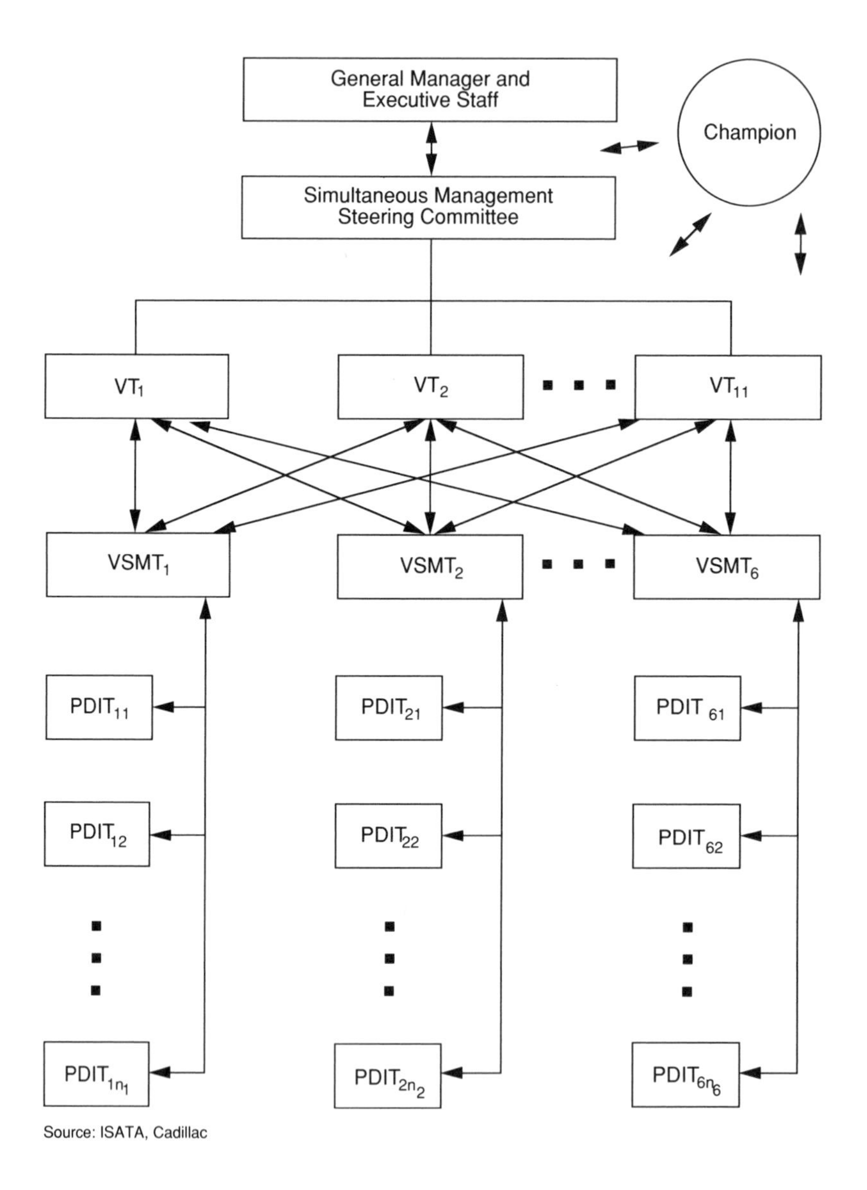

Source: ISATA, Cadillac

Figure 15-2. Large CE projects involving many different systems, such as those in the automobile industry, require many teams. In the system adopted by Cadillac, there is a strong hierarchy.

11. Vendors of major items of machinery and of major components should be brought into the project early. They should be selected from a number of companies before the concept is approved, and a member of the company should be brought into the task force straightaway. It can also be advantageous for vendors of manufacturing plant to be involved in continuous improvement programs, as is done at GM Engine Division, with one of their employees in the plant.

    In this way, bottlenecks and minor problems can be eliminated on a continuing basis. The vendor gains knowledge that will help build better machines for all customers from this close observation of its equipment, while the customer gains increased efficiency. Uptime in almost all U.S. plants is lower than it should be, and continuous improvement is the key to matching the Japanese.
12. When engineers from principal vendors of components and systems are involved in a major product, they should spend the critical design period working in the product design office. Of course, they will be in contact by telecommunications lines throughout the project so that they can exchange CAD drawings. But that is not enough when the design reaches a critical stage and close cooperation is needed. Likewise, vendors of manufacturing equipment should spend some time working alongside manufacturing engineers.
13. The chosen vendor should supply the prototypes, rather than a specialist prototype shop. In this way it gains familiarity with the product before it is faced with the problem of producing it in high volume, and it will be able to voice doubts about tricky aspects of the tooling well before the critical stage. Some vendors may not have suitable facilities, in which case they should control the manufacture of the prototypes and be responsible for quality and for learning

from the experience. They should be encouraged to either invest in prototype plant or set up a close relationship with a prototype specialist.

14. In each project, early verification of the basic design and genuine customer approval for the concept is required. It has been stressed many times that the customers' voice is the start of a project. It is also important that as soon as the customers' wants have been turned into a design, a clinic be held to show a mock-up or sketch to obtain verification.
15. The concept should not be released until an analysis of the production processes and costs has been undertaken, by DFMA or some other appropriate technique. This is not a job that can be shrugged off until later in the program; if it is, there will almost certainly be minor flaws causing major repercussions in costs and scrap rate.
16. Manufacturing should be given a budget to develop rough-and-ready test cells to try out new techniques. If facilities are not available in-house — and they should be — the vendor of the relevant machinery in the task force should do so. This budget should be separate from the normal development budget and should be related to each project.
17. One of the most difficult aspects, and one that can do more damage to a product than any other, is a sudden change in the market or among the competitors' products. How does the task force react? It should set guidelines. Where the addition of a feature is involved, it must recognize that if that feature has not already been developed, to attempt to add it can have one of two effects. Either

    - the product comes to market 6 to 12 months late, with the appropriate features but with its potential profitability substantially reduced, or
    - the product comes to market on time, but with the inadequately engineered and unreliable new item. The result

is reduced sales, increased service costs, and one step down the road to failure. This aspect of CE is discussed more fully in the next chapter.

18. The career paths for members of the task forces need to be maintained, and they need to know that this is so. Generally, they will spend a certain amount of time on the task force and then return to their departments.
19. Teamwork needs to be fostered, since it is the lifeblood of CE; petty political maneuvers should be positively discouraged.
20. Managers should be reminded daily that they need to foster change and completely wipe away the old, unsuccessful way of doing things.
21. Emphasis needs to be laid on business success. Whether or not the task force becomes a business center running a minibusiness, members need to be aware that profit from their product is the target. All actions they take should be aimed at making the business successful.

CHAPTER SIXTEEN

Surprises and Shorter Product Lives

Flexibility and discipline needed to handle changes in competition
The move to shorter lead time means renewing products more quickly — but how often?

Discipline is a vital ingredient in concurrent engineering, particularly when the market or competitive levels change suddenly. For example, halfway through a three-year new model program a competitor may introduce some new technology that makes your new model obsolete before it enters production. Or a shift in the demand pattern may lead to sales falling off faster than expected — say 12 months after introduction. In either case, panic is likely to erupt in some quarters. If CE is only in the fledgling stages, there will be enormous pressure to do something — anything that might boost sales in the short term.

This will be a major test for the new culture. In the old days, sales would start pressuring engineering to redesign the product, but now the decision is up to the CE team, which includes the sales representatives who agreed to the specification. If sales have fallen sharply, then errors must have been made in the handling of QFD, or the customer's preference has changed quickly — as happened in the energy crisis, when gas mileage suddenly began to sell cars.

An investigation is clearly needed, and it may be appropriate to modify the product in some way to overcome the weaknesses.

But the problems should not be allowed to alter the program. That is how companies turned programs into fire-fighting exercises in the past. For example, suppose competitor A makes air-conditioning a standard feature, and competitor B introduces a new model that is roomier than yours. In response, sales immediately requests that air-conditioning be added and seats be redesigned to increase rear legroom. The result is air-conditioning equipment that is poorly installed and seats that are less comfortable, at added cost. Thus, every customer has effectively been given a discount of $500 yet forced to pay a higher price for it.

With CE, that approach is quite unacceptable. There is no option but to go back to the customers and find out what they really want. Sales's view that air-conditioning as a standard feature will clinch the deal will be tested and quite likely found to be incorrect. Perhaps customers are turned away because of the latest paint job, some other fault no one is discussing, or sloppiness in the dealerships. QFD will give the team some clues, and with some brainstorming, they are likely to devise some way of overcoming the weakness. But whatever happens, the voice of the customers must be heard again.

Breakthrough in Technology

More serious is the situation in which a competitor introduces some startling new technology. One such case was when Canon introduced its color laser copier, a machine that not only produces copies of remarkable quality, but also gives the operator many editing options, such as changing the proportions of the original or changing the colors.

At the time, a major competitor was halfway through a program to develop a monochrome machine with similar features and similar prices — but no color. The task force saw at once that the machine would be uncompetitive, and that the standard trick of taking some cost out would not help. The simple fact was that the fundamental technology was missing. The project was abandoned.

How can such a situation be avoided? Unless a corporation can guarantee that it will always lead in technological development, a competitor will occasionally steal a march on it. One way of mitigating the effects of the situation, of course, is to increase investment in R&D, with long-term horizons.

Also, systems and components that are likely to come into use in the next few years should have been designed and developed, but without a specific product in mind. In the automobile industry, features such as full-time four-wheel drive, torque control to prevent wheelspin, four-wheel steering, active-ride suspension, four-valve cylinder heads, and CD-based navigation systems should be sitting on the shelf waiting to be produced.

In the personal computer industry, products that might be in the R&D arsenal waiting for the right moment include optical disk drives, parallel-processing nodes, RISC processors, graphics processors for input of images and ultra-rapid delivery to the screen, 486-based machines, and ultra-lightweight laptops. In both cases, the right moment is when the customers' voice says the product is needed, or when the competition forces the issue and the customers respond positively — but only then. The worst possible action is to rush a system into production simply because a competitor has done so. The customers may yawn and continue to buy the simpler, cheaper product.

Discipline, and the forethought that the task force approach fosters, will reduce surprises and make the corporation better able to respond to them than in the past. No more will it respond by adding undeveloped components to a properly proven product. Nor will it rehash a design after tooling has been completed. Instead, it will refocus the product so that it attacks a market where it will be competitive, and lay plans for an updated design that requires minimal tooling changes yet overcomes the weaknesses.

Of course, the disciplined approach is difficult to follow, especially when the chief executive is calling for immediate action, the sales people are stirring up trouble, and support for the new ideas suddenly seems thin on the ground. It is so much easier to start doing things — even the wrong things — to take people's minds off the real problem. But experienced executives know that panic action usually leads to more errors in judgment and in execution. Urgency is needed, but all the actions must be under control. Essential steps such as a manufacturing feasibility study are even more important with an urgent modification to a product than for a new concept because the tooling is already either ordered or in use.

Shorter Product Lives

Once these hurdles have been cleared, the CE approach will gain some maturity, and the chief executive will admit that the patience the task force showed under fire strengthened the corporation. As time goes on, the benefits will be apparent, and lead time will have fallen. It is then that the discussion of whether product lives should be shortened in accordance with the new shorter lead times becomes relevant.

For example, in the automobile industry the traditional lead time of five to seven years limited full model changes to six-to-eight year intervals. Cadillac used to take three years for a face-lift, so the pattern was for a new model every seven years, and a face-lift halfway through the model life. However, engineers started work on the face-lift almost as soon as the model entered production. Obviously, their first thoughts were to eliminate some of the drawbacks that they had seen in the development of the model rather than listen to the customers' voice. Customers had hardly had time to decide what they liked and did not like about the car.

Some mass-produced cars have lead times of three to four years, so models can be renewed every five years, while the Japanese rebody most of their cars at four-year intervals. So if the lead time is reduced to 24 months, does that mean every model is renewed at intervals of 30 to 36 months? Some engineers accept this as a natural progression, but it is not necessarily so.

Lead time and product life vary enormously according to the product. The Japanese anticipate lives of only 6 months for

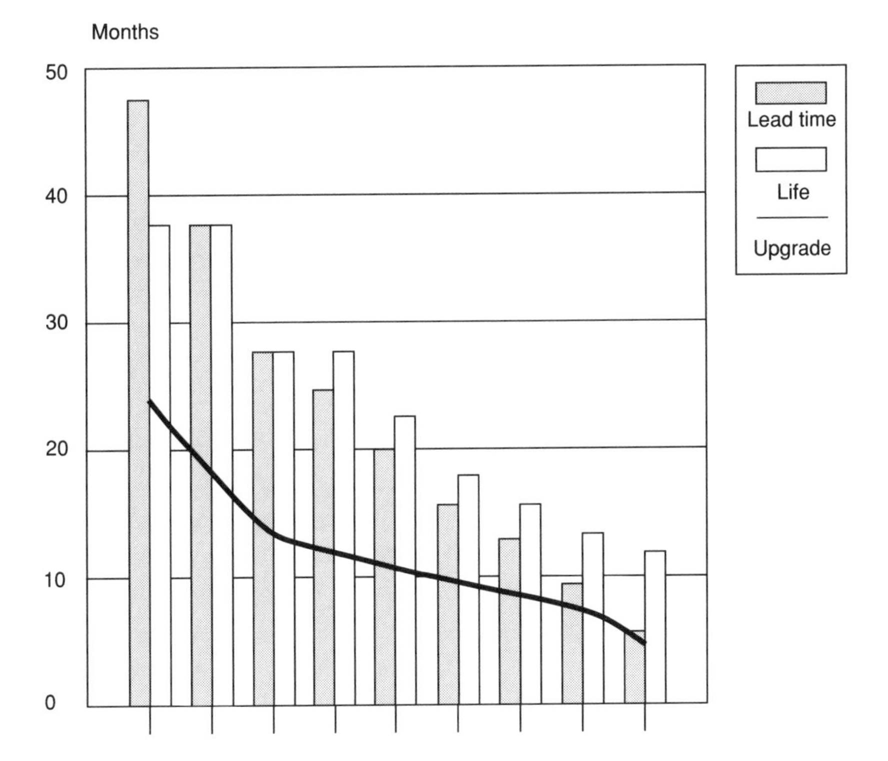

Figure 6-1. With over-the-fence engineering, lead time dictates the intervals at which new models are introduced and products are upgraded, as in the left-hand columns on the graph. As progress is made with CE, reduced lead times give management flexibility in reducing the lifespan of a product.

the cheaper personal audio equipment such as competitors to the Sony Walkman, but the more expensive units remain on the market for 12 to 18 months. The short lives resulted from extreme competition from the Japanese electronics companies, and those in Korea, Taiwan, and mainland China. The leaders were forced to introduce new models regularly to maintain market share.

However, every component on the new model is not redesigned. These units are based on a combination of standard chassis and mechanisms with standard ICs. The new models depend on extra features — such as a tuner — and different colors or shape, with the redesign of some of the elements. For example, rigid PCBs might be replaced with flexible units on one model, and a year later some smaller chips might be used. Then, from time to time, the manufacturer introduces new chassis, which are smaller than the previous ones. In addition, these machines are produced on flexible machinery, with robot insertion of some components, which can be reprogrammed to handle the new products.

New Model Every Day

In motorcycles, a similar situation arose in the mid-1980s, when Yamaha set out to catch Honda in sales volume. The result was a horrendous array of new model launches; at one point Honda was introducing a new model onto the Japanese market every day. Since in many cases the changes were cosmetic, the factories could just about cope with this number of model changes. Now, most Japanese motorcycles are equipped with body panels, so that the power unit and frame are less in evidence than in the past. Therefore, model changes can be made without altering the mechanical components every time. Although the number of model introductions is not as high as it was at the height of the battle between Honda and Yamaha, the product lives are still very short, and few in other industries would wish to contend with similar new-model introduction programs.

For many electronic components, the life is 12 to 24 months, and even for minicomputers, it is around 24 months. Again, machines are not completely redesigned at these intervals, but it is necessary to introduce derivatives frequently and to upgrade performance at regular intervals. Laptop computers are a typical field in which developments are being made at frightening speed. The impetus in laptop computers has been weight reduction. In 1988, a machine that weighed 15 pounds was competitive; now it needs to weigh less than 7 pounds, complete with hard disk drive. A machine introduced as state of the art in 1988 was seen gathering dust in a shop in 1990. In the interval, it had been completely outclassed in every respect by other machines, and production had been discontinued.

This trend is seen in many different markets, but not in every one. For example, the lives of mainframe computers and supercomputers are still long enough for manufacturers to introduce them to the market about 12 months before sales start, so as to accumulate orders. These are exceptions rather than the rule.

In practice, a combination of economics and market demand will tell the task force when they need to replace an existing model. However, some manufacturers take the view that they cannot reduce the life of the product beyond, say, five years because that length of time is needed to amortize the tooling. For example, Ford of Europe took considerable risks in deciding that its main models should remain in production for 8 to 10 years, with a major face-lift halfway through their lives. The first risk was that sales would fall during the last few years of the life of the product, and this appears to have been the case with the Sierra hatchback and Sapphire sedan. The second risk was that the project team would be working to improve such an old

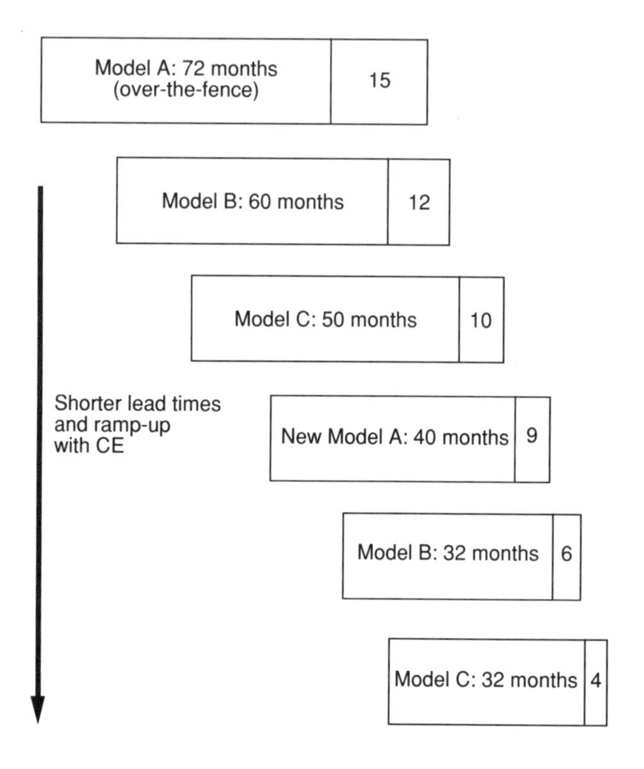

Figure 16-2. Progress with CE leads to shorter intervals before product introductions, requiring better management control.

design that their benchmarks would be far too low. This appears to be the problem with the new European Escort, although the situation was exacerbated somewhat by the lack of the expected new engines at the introduction of the models.

Shorter Product Lives Means Cheaper Tooling

The Japanese have shown that if the customer demands a new model every four years or every four months, it is possible to reduce the cost of tooling to maintain profitability. Any corporation seeking to reach world-class standards must be prepared to change its operations similarly. Competitiveness will be lost if investment criteria are allowed to limit response to the market. New techniques must be developed to allow the changes dictated by the market to be implemented without jeopardizing profitability.

Alternatively, it is necessary to carve out a niche as a company that abhors planned obsolescence and guarantees to keep products and replacement parts on hand for a certain time. In other words, a complete marketing package is built around the requirements of a certain sector of customers. Lip service to certain ideals will not do.

The customers, therefore, must be allowed to dictate how often models are changed. They may not want the model to be renewed very other year, especially if the product is mature, and may decide to buy a product that will be on the market longer. Or they may insist on buying the latest thing as soon as they can. Only they will know, and their immediate answer to any direct question about changes may not reveal what they really want. Undoubtedly, this is a question for QFD to answer.

The rhythm of an engineering department is often such that the engineering of the new program is starting as the old one finishes. Some engineers will be needed to work on derivatives, and others on improvements, leaving a nucleus to join the task force for the next model. However, if the lead time is halved, then the team could work on a completely different model in between succeeding models, so that when they came back to the old model their perspectives would be wider. Also, since the product had been in production for some time, useful data concerning its good and bad features could be obtained from all customers — including manufacturing and service personnel. Finally, with such a short lead time, there would be small chance of the customers' preferences changing or of the competitors springing surprises before production started.

Looking ahead a few years, an efficient CE organization with extensive CAD/CAM could engineer a new car model in 11 months, and the corporation could adopt a product life of 12 months. The team would start work almost as soon as the previous model hit the showroom. Therefore, it would be listening to customer demands that had become obsolete. This misreading of customer demands is not so different from that which occurs when a company starts a project five years before production starts and has difficulty in obtaining up-to-date information — except that in the former case work is started so much nearer the introduction date.

In both situations, the combination of a multidisciplinary team and CE gives the corporation a head start. With the diligent use of QFD, the team will find out just what it has to do to satisfy the customer. In other words, the more competitive business becomes, the more essential concurrent engineering becomes.

CHAPTER SEVENTEEN

Future Concurrent Engineering

New grouping of departments
Business responsibility in task force
Common data base for all departments
Restructuring in industry

Once a company adopts concurrent engineering wholeheartedly, its culture must and will change. The old management structure is likely to prove inadequate in some areas, and key activities will be more concentrated than previously. People working on new projects will be more highly motivated than in the past, so that the gains inherent in CE can be compounded. However, unless the attitude of managers in the company toward design is changed, the full benefits will not be realized.

The future of CE in any company has no arbitrary limits, no restraints imposed from outside. What CE becomes depends solely on the flair and imagination of managers in the company. And imagination will be needed because the system of the task forces will be opposed by some middle managers who see loss of status or their territory being usurped by the teams. Others will worry about the changes taking place in the structure of the company. Senior managers need to keep their fingers on the pulse of their company and to anticipate the directions in which CE is taking them, and where they wish to guide it. They should resist the thought that the task forces are going too far, and that the company is changing from a well-structured and disciplined force to a mass of people being pulled along willy-nilly by these teams. Instead, they need to learn to live with less direct control, but with better real control of projects. Then, they need to prevent the structure maniacs in the organization from demanding that the task forces fall into the normal line of reporting. Each request of that type begs a simple answer: "Show how that would improve our new product introduction and profitability."

If the complaint cannot be turned into a method of improving either profitability or new product introduction, it is invalid. Greater control of the deployment of personnel, for example, is not a valid reason for making changes. This single-minded approach to business is fundamental to CE, and to success in the 1990s.

Product Design and Manufacturing Regrouped

In most companies there will be a regrouping of the main departments, with product design and manufacturing engineering coming under the control of one vice president. The natural counterweight will be the vice president responsible for marketing, sales, and service. Finance and personnel are likely to wield less power than in the past, although it is important that the board be balanced.

Where product design and manufacturing engineering come under one vice president, new divisions responsible for the main subassemblies will emerge. For example, in vehicle manufacture, power-train design and manufacture is likely to form one division, and body and trim another. In electronics, one division might handle printed circuit boards and their assembly, another ICs and components, another peripherals, and the last housings. In all cases, however, the division would handle design, development, and manufacture.

It is important that each division be fully responsible for all its activities. For example, chassis, electrical, and trim might be put together because they are all assembled to the vehicle on the final assembly line. If they were separated, advances in automation might be difficult to implement.

In changing these relationships, many companies may find it beneficial to either reorganize their existing buildings or erect new ones that reflect the core position of task forces. Offices for task forces could be at the center of a new building, with all the departments around them, or they could be housed in a building serving as a bridge between R&D and other departments.

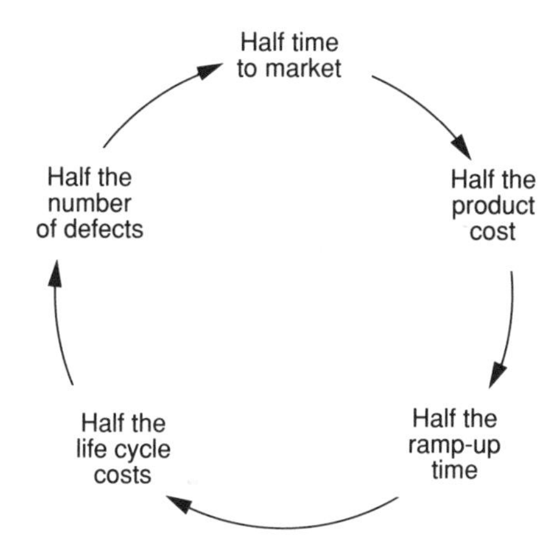

Phase 0 Requirements	Phase 1 Concept	Phase 2 Detail	Phase 3 Quality	Phase 4 Ramp-up

Source : Digital Equipment

Figure 17-1. Once concurrent engineering is established, the goals are continuous halving of lead times, product costs, and lifecycle costs.

Profit Center?

The next area of potential conflict is the range of responsibilities given to members of the task force. Their responsibility may end when Job One goes down the line, or when the production rate reaches the planned level; or the task force may remain in existence throughout the life of the product.

To be effective, the team needs to remain intact at least until production has been ramped up, and be responsible for that ramp-up. Only then is the company likely to achieve full production and reach the break-even point quickly.

In the future more teams will continue to operate throughout the life of the product so that they can control minor changes, service problems, and monitor customer requirements continuously. In many companies the task force will develop into a profit center with total business responsibility for the product. This is the logical result of CE and the task force approach.

However, if the task force becomes a self-sufficient profit center, the company structure will change; the departments that now handle mainstream activities may instead provide services to the task forces only. Clearly, senior managers need to be ready for this eventuality before it arises. They also need to have a flexible approach to the desired lines of command and to the changes in career structures that will need to be implemented to bring about that end.

Whether this approach is viable will depend principally on the size of the company and the number of mainline products it has. With 4 or 5 main products, each divided into 4 subunits, the management structure would form a practical management pyramid; but with 15 or 20 main products, the system could be unmanageable.

There are several dangers in the implementation of CE. With an unimaginative approach, design could become too formal, relying heavily on standards and specific techniques and not enough on the imagination of designers. To overcome this problem, managers will need to allow enough room for initiative and to encourage brainstorming to solve problems. The balance within a given task force, among various task forces, and between the task forces and the departments will need to be guided carefully. Initial attitudes by people outside task forces that the teams are insignificant can easily give way to the attitude that task forces hold all the power. Again, good management is needed to maintain the correct balance — and the preferred balance will vary from company to company.

Common Data Base

While companies are gaining their first experiences with CE, technical developments and changes in the environment and in the market will continue to move the goalposts. The increasing power of CAD workstations will allow more and faster simulations and improved results, but object-oriented data bases with ultrafast access will make distributed systems more user-friendly. Ultimately, there will be a common relational data base available for all departments. Product data will be maintained by product design, but they will not be the property of one department. Instead, the common data will be loaded into the workstations in any department, where they will be converted into the form required.

In manufacturing, the data will be suitable for generating cutter paths and basic dimensions for fixtures or die forms. Marketing will see the styling drawings and outline of components with the specifications they require to maximize sales. Service will be able to find the location of key areas such as sockets and fillers. Data supplied to finance will be available in such a way that costings can be made easily.

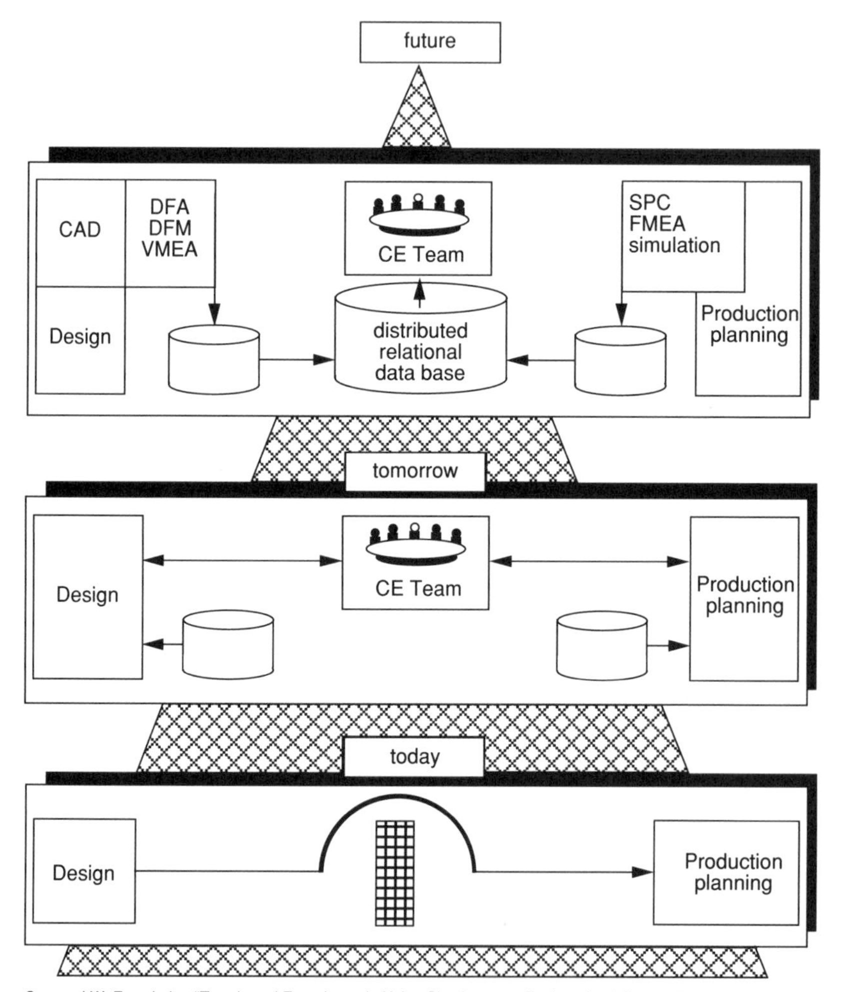

Source: I.W. Eversheim, "Trends and Experience in Using Simultaneous Engineering," *Proceedings of the 1st International Conference on Simultaneous Engineering*, London, December 1990, p. 23.

Figure 17-2. In the future, automated design modules and a distributed relational data base will be the core tools in CE.

Thus, each department will be able to see the latest data and work on them from concept to final design, and will be forewarned of any difficulties. Of course, if any department wants a change, it will contact its member of the task force — not product design — to make proposals for the changes it needs. The other members of the task force will be able to run simulations and then approve or modify the three-dimensional model as appropriate.

Also, the increased number of simulation packages available will allow many more aspects of design to be simulated than at present. Data logged for SPC purposes will provide an indication of what can be achieved with future products, so that tolerancing will be much nearer the mark than at present. Not only will this software, and the more powerful computers in use, allow simulation to be done much more quickly than in the past, but it will allow the accuracy of the work to be improved continually — and the full benefits of these gains in hardware and software will come only with concurrent engineering.

It follows that once task forces are fully operational, and they and the backup departments are using such software, the speed of product development and the productivity of design and production engineers will increase substantially. This will lead to much smaller design departments or much larger model ranges. Where product lead time is now being reduced from 48 to 36 months with CE, continuous lowering of targets will lead to a period of 20 to 28 months within five years, and later on to 12 to 18 months.

Vendors in the Group

Meanwhile, vendors are likely to be brought more closely into the net, relying on the same data base as the customer to speed up design work. Confidentiality of data is always a cause for concern, and the transmission of proprietary data between departments and to vendors requires considerable attention to security, with the use of passwords, ID cards, and encryption for external transmissions. In this respect, the new "smart" cards that incorporate microprocessors are a distinct advance.

Because of their involvement in task forces, some vendors may find themselves working for one major manufacturer only in each industry. Others may find that to serve several different companies they need to set up separate companies, each situated near a major client. However they react, vendors will work much more closely with their customers, and this could lead to a regrouping of industry, with some vendors becoming part of the group headed by their main customer.

Therefore, in the future CE will not just penetrate deeper into the companies that use it, but will change the fundamental structures of those companies, and of the industry sectors.

Bibliography

Books

Ackoff, R. L. *The Art of Problem Solving.* New York: Wiley-Interscience, 1978.

Akao, Y., ed. *Quality Function Deployment: Integrating Customer Requirements into Product Design.* Cambridge, Mass.: Productivity Press, 1990.

Andreasen, M. M., S. Kahler, and T. Lund. *Design for Assembly.* Kempston, England: IFS Publications/Springer Verlag, 1988.

Bendell, A., J. Disney, and W.A. Pridmore, eds. *Taguchi Methods: Applications in World Industry.* Kempston, England: IFS Publications/Springer Verlag, 1989.

Clark, Kim B., and Takahiro Fujimoto. *Product Development Performance: Strategy, Organization, and Management in the Automobile Industry.* Boston: Harvard Business School, 1991.

Cullen, J., and J. Hollingum. *Implementing Total Quality.* Kempston, England: IFS Publications, 1987.

Fukuda, R. *Managerial Engineering: Techniques for Improving Quality and Productivity in the Workplace.* Rev. ed. Cambridge, Mass.: Productivity Press, 1986.

Grant, Eugene L., and Richard S. Leavenworth. *Statistical Quality Control.* New York: McGraw-Hill, 1988.

Ishikawa, Kaoru. *What Is Total Quality Control? The Japanese Way.* (Translated by David J. Lu.) Englewood Cliffs, NJ: Prentice-Hall, 1985.

Kivenko, K. *Quality Control for Management.* Englewood Cliffs, NJ: Prentice-Hall, 1984.

Kobayashi, I. *20 Keys to Workplace Improvement.* Cambridge, Mass.: Productivity Press, 1990.

Miller, D. *The Icarus Paradox.* New York: Harper Collins, 1991.

Nakajima, S. *TPM Development Program: Implementing Total Productive Maintenance.* Cambridge, Mass.: Productivity Press, 1989.

Oakland, J. S. *Statistical Process Control.* London: Heinemann, 1986.

———. *Total Quality Management.* London: Heinemann, 1989.

Pascale, Richard T. *Managing on the Edge: How the Smartest Companies Use Conflict to Stay Ahead.* New York: Simon and Schuster, 1990.

Shingo, S. *A Study of the Toyota Production System from an Industrial Engineering Viewpoint.* (Rev. ed.). Cambridge, Mass.: Productivity Press, 1989.

Taguchi, Genichi. *Introduction to Quality Engineering.* Tokyo: Asian Productivity Organization, 1986.

Wilson, G., R. M. G. Millar, and A. Bendall. *Taguchi Methodology with Total Quality.* Kempston, England: IFS Publications, 1990.

Conference Papers

Adams, R. M., and M. D. Gavoor. *Implementing a Companywide Quality Strategy.* SAE Technical Paper 880328m, 1988.

Aswad, A. A., and J. W. Knight. "Comparative Aspects of QFD and Simultaneous Engineering." In Vol. 1 of *Proceedings of the 21st International Symposium on Automotive Technology & Automation (ISATA)*, Wiesbaden, November 1989.

Barron, D.D. *Simultaneous Engineering at Delco Remy*. SME Technical Paper MM88-154, 1988.

Broughton, T. "Simultaneous Engineering in Aero Gas Turbine Design and Manufacture." In *Proceedings of the 1st International Conference on Simultaneous Engineering*, London, December 1990.

Chelsom, J. "Developing Partnerships for Simultaneous Engineering." In Vol. 1 of *Proceedings of ISATA*. *See* Aswad and Knight.

Corbette, J. "Reduction of Engineering Timescales Using Simultaneous Engineering." In *Proceedings of Auto Tech*, Birmingham, England, November 1989.

de Lorge, D. "Role of the Equipment Supplier in Simultaneous Engineering." In *Proceedings of the 1st International Conference on Simultaneous Engineering*. *See* Broughton.

Eigenfeld, B., B. Kroeger, and T. Ortleb. *3D CAD/CAM Design of a 4-valve Cylinder Aluminum Head*. SAE Technical Paper 900655, 1990.

Eversheim, I. W. "Trends and Experience in Using Simultaneous Engineering." In *Proceedings of the 1st International Conference on Simultaneous Engineering*. *See* Broughton.

Ford, J. C. "Simultaneous Engineering (Design to Manufacture)." In *Proceedings of Auto Tech*. *See* Corbette.

Gilroy, T. "People and Organisation in Managing Simultaneous Engineering." In *Proceedings of the 1st International Conference on Simultaneous Engineering*. *See* Broughton.

Grant, D. "Simultaneous Engineering Applied to Data Communications Products." In *Proceedings of the 1st International Conference on Simultaneous Engineering*. *See* Broughton.

Hampson, R., and K. Foxley. "Simultaneous Engineering in Nissan's European Operations and Implications for Suppliers." In *Proceedings of the 1st International Conference on Simultaneous Engineering. See* Broughton.

Hinkley, J. P., Jr. *Early Designs for Manufacturing Quality.* SME Technical Paper MM88-151, 1988.

King, J. A. "Simultaneous Engineering: How to Make It Work." In *Proceedings of Auto Tech. See* Corbett.

Liesgang, G. "Life-span Oriented Simultaneous Engineering." In Vol. 1 of *Proceedings of ISATA. See* Aswad and Knight.

MacDow, R. W. "The Technology of Simultaneous Engineering." In Vol. 1 of *Proceedings of ISATA. See* Aswad and Knight.

Nichols, K. "Competing through Design: Today's Challenge." In *Proceedings of the 1st International Conference on Simultaneous Engineering. See* Broughton.

Ohara, M. "CAD/CAM at Toyota Motor Corporation." *Japan Annual Reviews in Electronics, Computers & Telecommunciations* 18 (no. 3: 2).

Pao, T.W., M.S. Phadke, and C.S. Sherrerd. *Computer Response Time Optimization Using Orthogonal Array Experiments.* 1985 IEEE International Conference on Communications, June 23-26.

Partishkoff, D. J. *Early Sourcing and Involvement of Steel Companies on the 1989 Cadillac DeVille and Fleetwood Programs.* SAE Technical Paper 890342, 1989.

Ritter, W. "Simultaneous Engineering: An Organizational Prerequisite for Efficient and Rapid Technology Innovation." In *Proceedings of the 1st International Conference on Simultaneous Engineering. See* Broughton.

Ruding, G. "Meeting the Market Demand with Simultaneous Engineering." In *Proceedings of the 1st International Conference on Simultaneous Engineering. See* Broughton.

St. Charles, D. P. "Part Interchangeability, Tolerancing and Manufacturing Cost." In Vol. 1 of *Proceedings of ISATA. See* Aswad and Knight.

Schoeffler, G. H. "A Supplier's Approach Towards Reduction of Product Development Time in Automotive Lighting by Simultaneous Engineering." In Vol. 1 of *Proceedings of ISATA. See* Aswad and Knight.

Schonwald, B. "Simultaneous Engineering and Its Part in Research and Development." In Vol. 1 of *Proceedings of ISATA. See* Aswad and Knight.

Walklet, R. H. "Simultaneous Engineering: A Cadillac Perspective." In Vol 1 of *Proceedings of ISATA. See* Aswad and Knight.

Wallace, J. "Simultaneous Engineering from an Engine Manufacturer's Viewpoint." In *Proceedings of the 1st International Conference on Simultaneous Engineering. See* Broughton.

Warr, R. E. *Failure Modes and Effects Analysis Method for New Product Introductions.* SAE Technical Paper 841600, 1984.

Worreschk, W. "Simultaneous Engineering of a New Multi-link Rear Suspension." In Vol. 1 of *Proceedings of ISATA. See* Aswad and Knight.

Wulf, I. A., and R. Sterbl. "Simultaneous Engineering for the Automotive Industry Using Workstations and Computer Servers." In *Proceedings of Auto Tech. See* Corbette.

Yamazoe, T. "Simultaneous Engineering: A Nissan Approach." In *Proceedings of the 1st International Conference on Simultaneous Engineering. See* Broughton.

Young, C. *Simultaneous Engineering: A Break with Tradition.* Birmingham, England: The Perfect Partnership, May 1989.

About the Author

John R. Hartley, well known internationally as an engineering writer, traveled extensively and drew on his considerable experience in researching and writing this book. He spent much time visiting American, British, German, Italian, and Japanese companies and studying concurrent engineering. His work as a design engineer in the automobile industry and more than twenty years as an engineering writer and analyst, including nine years in Japan — he first wrote about the Toyota Production System and just-in-time in 1977 — all came into play when he studied concurrent engineering, about which he is as enthusiastic as any of the experts he met.

He has written books on robots, flexible manufacturing systems, flexible automation in Japan, the need to adopt Japanese manufacturing systems, and management of vehicle manufacture.

Index

OTHER BOOKS FROM PRODUCTIVITY PRESS

Productivity Press publishes and distributes materials on continuous improvement in productivity, quality, and the creative involvement of all employees. Many of our products are direct source materials from Japan that have been translated into English for the first time and are available exclusively from Productivity. Supplemental products and services include membership groups, conferences, seminars, in-house training and consulting, audio-visual training programs, and industrial study missions. Call toll-free 1-800-394-6868 for our free catalog.

The Benchmarking Management Guide

American Productivity & Quality Center

If you're planning, organizing, or actually undertaking a benchmarking program, you need the most authoritative source of information to help you get started and to manage the process all the way through. Written expressly for managers of benchmarking projects by the APQC's renowned International Benchmarking Clearinghouse, this guide provides exclusive information from members who have already paved the way. It includes information on training courses and ways to apply Baldrige, Deming, and ISO 9000 criteria for internal assessment, and has a complete bibliography of benchmarking literature.
ISBN 1-56327-045-5 / 260 pages / $39.95 / Order BMG-B204

The Benchmarking Workbook
Adapting Best Practices for Performance Improvement

Gregory H. Watson

Managers today need benchmarking to anticipate trends and maintain competitive advantage. This practical workbook shows you how to do your own benchmarking study. Watson's discussion includes a case study that takes you through each step of the benchmarking process, raises thought-provoking questions, and provides examples of how to use forms for a benchmarking study.
ISBN 1-56327-033-1 / 169 pages / $29.95 / Order BENCHW-B204

Cycle Time Management
The Fast Track to Time-Based Productivity Improvement
Patrick Northey and Nigel Southway

As much as 90 percent of the operational activities in a traditional plant are nonessential or pure waste. This book presents a proven methodology for eliminating this waste within 24 to 30 months by measuring productivity in terms of time instead of revenue or people. CTM is a cohesive management strategy that integrates just-in-time (JIT) production, computer integrated manufacturing (CIM), and total quality control (TQC). From this succinct, highly-focused book, you'll learn what CTM is, how to implement it, and how to manage it.
ISBN 1-56327-015-3 / 208 pages / $29.95 / Order CYCLE-B204

Design Team Revolution
How to Cut Lead Times in Half and Double Your Productivity
Kenichi Sekine and Keisuke Arai

This book addresses the problem of waste in the product design process. It shows how to apply continuous improvement methods that have been successful on the shop floor to the design function, to eliminate waste in time, materials, transport, and other areas. The authors show design managers how to establish one-piece flow, U-cell processes, and parallel design for design teams, and discuss the role of running changes, site diagnosis and machine inspection, and designer training.
ISBN 1-56327-008-0 / 328 pages / $85.00 / Order DTREV-B204

Fast Focus on TQM
A Concise Guide to Companywide Learning
Derm Barrett

Finally, here's one source for all your TQM questions. Compiled in this concise, easy-to-read handbook are definitions and detailed explanations of over 300 key terms used in TQM. Organized in a simple alphabetical glossary form, the book can be used either as a primer for anyone being introduced to TQM or as a complete reference guide. It helps to align teams, departments, or entire organizations in a common understanding and use of TQM terminology. For anyone entering or currently involved in TQM, this is one resource you must have.
ISBN 1-56327-049-8 / 208 pages / $19.95 / Order FAST-B204

Function Analysis
Systematic Improvement of Quality and Performance
Kaneo Akiyama

Function Analysis is a systematic technique for isolating and analyzing various functions in order to better design and improve products. This book gives you a solid understanding of Function Analysis as a tool for system innovation and improvement; it helps you design your products and systems for improved manufacturability and quality. It describes how Function Analysis can be used in the office as well as on the shop floor.
ISBN 0-915299-81-X / 269 pages / $60.00 / Order FA-B204

Handbook for Productivity Measurement and Improvement
William F. Christopher and Carl G. Thor, eds.

An unparalleled resource! In over 100 chapters, nearly 80 front-runners in the quality movement reveal the evolving theory and specific practices of world-class organizations. Spanning a wide variety of industries and business sectors, they discuss quality and productivity in manufacturing, service industries, profit centers, administration, nonprofit and government institutions, health care and education. Contributors include Robert C. Camp, Peter F. Drucker, Jay W. Forrester, Joseph M. Juran, Robert S. Kaplan, John W. Kendrick, Yasuhiro Monden, and Lester C. Thurow. Comprehensive in scope and organized for easy reference, this compendium belongs in every company and academic institution concerned with business and industrial viability.
ISBN 1-56327-007-2 / 1344 pages / $90.00 / Order HPM-B204

The Hunters and the Hunted
A Non-Linear Solution for American Industry
James B. Swartz

Because our competitive environment changes so rapidly — weekly, even daily — if you want to survive, you have to stay on top of those changes. Otherwise, you become prey to your competitors. Hunters continuously change and learn; anyone who doesn't becomes the hunted and sooner or later will be devoured. This unusual non-fiction novel provides a veritable crash course in continuous transformation. It offers lessons from real-life companies and introduces many industrial gurus as characters, as well providing a riveting story of two strong people struggling to turn their company around. The Hunters and the Hunted doesn't simply tell you how to change; it puts you inside the change process itself.
ISBN 1-56327-043-9 / 582 pages / $45.00 / Order HUNT-B204

Measuring, Managing, and Maximizing Performance

Will Kaydos

You do not need to be an exceptionally skilled technician or inspirational leader to improve your company's quality and productivity. In non-technical, jargon-free, practical terms this book details the entire process of improving performance, from why and how the improvement process work to what must be done to begin and to sustain continuous improvement of performance. Special emphasis is given to the role that performance measurement plays in identifying problems and opportunities.
ISBN 0-915299-98-4 / 284 pages / $39.95 / Order MMMP-B204

A New American TQM
Four Practical Revolutions in Management

Shoji Shiba, Alan Graham, and David Walden

For TQM to succeed in America, you need to create an American-style "learning organization" with the full commitment and understanding of senior managers and executives. Written expressly for this audience, A New American TQM offers a comprehensive and detailed explanation of TQM and how to implement it, based on courses taught at MIT's Sloan School of Management and the Center for Quality Management, a consortium of American hi-tech companies. Full of case studies and amply illustrated, the book examines major quality tools and how they are being used by the most progressive American companies today.
ISBN 1-56327-032-3 / 606 pages / $49.95 / Order NATQM-B204

The New Standardization
Keystone of Continuous Improvement in Manufacturing

Shigehiro Nakamura

In an era of continuous improvement and ISO 9000, quality is not an option but a requirement and you can't set or meet criteria for quality without standardization. Standardization lets you share information about the best ways to do things so that they will be done that way consistently. This book shows how to make standardization a living system of just-in-time information that delivers exactly the information that's needed, exactly when it is needed, and exactly where it is needed. It's the only way to sustain the results of your improvement efforts in every area of your company.
ISBN 1-56327-039-0 / 286 pages / $75.00 / Order STAND-B204

Quality Function Deployment
Integrating Customer Requirements into Product Design
Yoji Akao (ed.)

Written by the creator of QFD, this book provides direct source material on one of the essential tools for world class manufacturing. More and more companies are using QFD to identify customer requirements, translate them into quantified quality characteristics, and then build them into their products and services. This casebook introduces the concept of quality deployment as it has been applied in a variety of industries in Japan.
ISBN 0-915299-41-0 / 387 pages / $85.00 / Order QFD-B204

Software and the Agile Manufacturer
Computer Systems and World Class Manufacturing
Brian Maskell

The term "agile manufacturing" describes responsive, flexible manufacturing that can deliver better products, faster, at lower cost. This book is the first to address the critical question of how computerization can aid the transition. It shows how computer systems and software designed for individual departments or functions can be adapted to create a world class manufacturing environment that's integrated companywide. Case studies reveal the common characteristics companies have shared in the challenge to computerize and provide guidelines for companies just starting out. This is a non-technical, practical guide.
ISBN 1-56327-046-3 / 424 pages / $49.95 / Order SOFT-B204

The Teamwork Advantage
An Inside Look at Japanese Product and Technology Development
Jeffrey L. Funk

How are so many Japanese manufacturing firms shortening product time-to-market, reducing costs, and improving quality? The answer is teamwork. Dr. Funk spent 18 months as a visiting engineer at Mitsubishi and Yokogawa Hokushin Electric and knows firsthand how Japanese corporate culture promotes effective teamwork in production, design, and technology development. Here's a penetrating case study and analysis that presents a truly viable model for the West.
ISBN 0-915299-69-0 / 508 pages / $49.95 / Order TEAMAD-B204

TQM for Technical Groups
Total Quality Principles for Product Development
Kiyoshi Uchimaru, Susumu Okamoto, and Bunteru Kurahara

Achieving total quality in product design and development is a daunting but essential goal for technical personnel. This unprecedented and highly practical book was written especially for technical groups working to achieve total quality in product development. Originally published by JUSE, the Union of Japanese Scientists and Engineers, the book includes an important case study of NEC IC Microcomputer Systems, winner of the Deming Prize. A separate section of the book addresses all the changes required in corporate management to institute TQM at the product design level. Step-by-step instructions, with specific examples of each, show you how to plan, implement, and sustain an effective TQM program.
ISBN 1-56327-005-6 / 258 pages / $60.00 / Order TQMTG-B204

The Unshackled Organization
Facing the Challenge of Unpredictability Through Spontaneous Reorganization
Jeffrey Goldstein

Managers should not necessarily try to solve all the internal problems within their organizations; intervention may help in the short term, but in the long run may inhibit true problem-solving change from taking place. And change is the real goal. Through change comes real hope for improvement. Goldstein explores how change happens within an organization using some of the most leading-edge scientific and social theories about change and reveals that only through "self organization" can natural, lasting change occur. This book is a pragmatic guide for managers, executives, consultants, and other change agents.
ISBN 1-56327-048-X / 208 pages / $25.00 / Order UO-B204

TO ORDER: Write, phone, or fax Productivity Press, Dept. BK, 541 NE 20th Ave., Portland, OR 97232, phone 1-800-394-6868, fax 1-800-394-6286. Send check or charge to your credit card (American Express, Visa, MasterCard accepted). **U.S. ORDERS:** Add $5 shipping for first book, $2 each additional for UPS surface delivery. We offer attractive quantity discounts for bulk purchases of individual titles; call for more information. **INTERNATIONAL ORDERS:** Write, phone, or fax for quote and indicate shipping method desired. For international callers, telephone number is 503-235-0600 and fax number is 503-235-0909. Prepayment in U.S. dollars must accompany your order (checks must be drawn on U.S. banks). When quote is returned with payment, your order will be shipped promptly by the method requested.

NOTE: Prices are in U.S. dollars and are subject to change without notice.

Productivity Press, Inc., Dept. BK, P.O. Box 13390, Portland, OR 97213-0390
Telephone: 1-800-394-6868 Fax: 1-800-394-6286